551
Fonteneau, Gilles.
Murmurs from the deep : scientific adventure in the Caribbean
New York : Arcade Pub., c2006.

SANTA MARIA PUBLIC LIBRARY

Discarded by
Santa Maria Library

MURMURS FROM THE DEEP

MURMURS FROM THE DEEP

SCIENTIFIC ADVENTURE IN THE CARIBBEAN

Gilles Fonteneau

Translated from the French by George Holoch

ARCADE PUBLISHING
NEW YORK

Copyright © 2005 by Gilles Fonteneau
Translation copyright © 2006 by Arcade Publishing, Inc.

All rights reserved. No part of this book may be reproduced in any form or by any electronic or mechanical means, including information storage and retrieval systems, without permission in writing from the publisher, except by a reviewer who may quote brief passages in a review.

FIRST ENGLISH-LANGUAGE EDITION

First published in French as *Les murmures du silence* by Editions L'Ancre de Marine

Library of Congress Cataloging-in-Publication Data

Fonteneau, Gilles.
[Murmures du silence. English]
Murmurs from the deep : scientific adventure in the Caribbean / by Gilles Fonteneau ; preface by François Bellec. — 1st English-language ed.
p. cm.
ISBN 1-55970-776-3
1. Oceanography — Caribbean Sea. 2. Underwater exploration — Caribbean Sea. 3. Plate tectonics — Caribbean Sea. 4. Fishes — Physiology. 5. Animal communication. 6. Fonteneau, Gilles. I. Title.

GC531.F66 2006
551.46'1365—dc22 2005010313

Published in the United States by Arcade Publishing, Inc., New York
Distributed by Time Warner Book Group

Visit our Web site at www.arcadepub.com

10 9 8 7 6 5 4 3 2 1

Designed by API

EB

PRINTED IN THE UNITED STATES OF AMERICA

Oh, the continual drunken Diversity of flights
and departures!
Eternal soul of navigators and their navigations!

— Fernando Pessoa, *Maritime Ode*
(tr. Edwin Honig and Susan M. Brown)

Contents

Preface

The theory of continental drift, which the Austrian Hans Suess intuited in the late nineteenth century and the German geophysicist Alfred Wegener developed further in his *Die Entstehung der Kontinente und Ozeane* in 1915, was almost immediately abandoned as an extravagant idea. But its truth had to be accepted in the 1960s, when systematic study of the paleomagnetism of the ocean depths began, and the principle of plate tectonics was articulated in 1967. Driven by the viscous movement of the magma, the continents have been drifting for the last 200 million years, as though they were floating on enormous rafts. In the area close to Central America, approaching the Pacific, the Caribbean plate, squeezed between the North and South American plates, strikes the Cocos plate, which is itself caught between the Nazca and Pacific plates. The slow, ponderous kinematics of the lithosphere, a broken eggshell, produces powerful and silent confrontations between tectonic plates. They erected the Andes and Rocky Mountain ranges, created the San Andreas Fault, and scoured out the great ocean trenches. They are the cause of the destructive earthquakes shaking the western edge of the Americas from Concepción to San Francisco and of the volcanic eruptions of the West Indies. The magma is breathing beneath our feet, and it is continually pushed up by geothermal forces along the rift of the Mid-Atlantic Ridge. Under its pressure, the Americas and Eurasia are moving apart by four centimeters each year. The Atlantic is twenty meters wider than at the time of Columbus.

For the last fifteen years, the Global Positioning System (GPS) has made it possible to pinpoint within a few meters, or even centimeters, depending on how it is used, any point on earth with reference to a universal structure established by a network of artificial satellites, and hence to measure one's movement from that reference point. In March 2002, Gilles Fonteneau anchored at Las Aves on a scientific mission to lay down some of these sensors. Absent from ordinary maps, Las Aves is a minuscule emanation of the Aves Ridge in the heart of the Caribbean Sea. One of the modest visible signs of the formidable viscous movement of continental drift, it contains barely enough space for the point of a compass. And that was precisely why a touchy group of Venezuelan soldiers ordered him to sail off. For this minuscule island is coveted, as are, in other oceans, Clipperton, Christmas, Bouvet, Desventurados, and Chesterfield, because a huge imaginary line traces around each of these points a two-hundred-mile zone of exclusive economic jurisdiction, subject to further legal extension if there is geographic continuity with another possession or the national territory of the sovereign state. And when the entire region produces oil, the landing of innocent scientists can look like a scene from a James Bond movie.

Captain Charcot, nicknamed "the Polar Gentleman" by Sir Ernest Shackleton was also drawn away from his professional career because he was a sailor at heart and a scientist by inclination. The great Antarctic explorer would no doubt likewise have called Gilles Fonteneau a "Caribbean Gentleman," with no reservations, as a recognition of quality. To a reader who may be wondering about the value of a few geodesic observations carried out from a little thirteen-meter catamaran when the scientific community has scattered intelligent sensors through earth and space, I offer the following significant fact: although plate tectonics is now an exact science, geophysicists are still unable to predict disturbances in the earth's crust. By a sinister coincidence — the very one that unleashes the unforeseeable character disorders of our planet — at the very moment that I was writing these lines on December 26, 2004, in Paris, a severe earthquake of magnitude 9, with its epicenter in the ocean northwest of Sumatra, produced a tsunami sweeping across the Andaman and Nicobar islands, plunging into the Bay of Bengal, and causing several hundred thousand deaths from Sri Lanka to Somalia. The Eurasian and Indo-Australian plates had suddenly resumed their ferocious antagonism while shifting the Sumatra plate by a dozen meters. Even though we do not know enough to make short-term predictions, the preservation of human life throughout the West Indies obviously depends on preventive observation of the variations in the plates' geophysical parameters, to which the *Prince de Vendée* expedition was intended to contribute. The Sumatra quake dramatically validates these efforts.

Gilles Fonteneau could be an avatar of a dozen practical adventurers periodically reincarnated in the course of centuries. Certainly of Amerigo Vespucci, a commercial agent for the

Medicis — who was good enough as a navigator in the Iberian voyages of investigation of the New World following Columbus to have been appointed chief pilot of the Casa de Contratación of Seville; of Jacques-Yves Cousteau, who made his little *Calypso* into a mythical vehicle for popular knowledge of the ocean; of Haroun Tazieff, the brilliant volcanologist; of Captain Charcot, president of the Yacht Club of France and self-taught explorer of the Arctic and the Antarctic on board the *Pourquoi-Pas?* Probably Gilles's *Prince de Vendée* can claim descent above all from Albert I of Monaco, the prince-navigator, who in the era of yachting at Spitsbergen, Jules Verne, and the birth of oceanography transformed his *Hirondelle* and *Princesse Alice I* and *II* into laboratories and encouraged his rich friends to do likewise in the name of science instead of wasting their time in social frivolity. Gilles Fonteneau might simply be a descendant of his namesake, the mysterious Jean Fonteneau. Jean Alfonse, in the Portuguese spelling, also known as Alfonse de Saintonge, was a well-known seaman in La Rochelle in the middle of the sixteenth century, but he was also known outside its walls: a pilot for Jacques Cartier in the voyage to Newfoundland, he was one of the first to transmit the Portuguese nautical tradition beyond the Iberian Peninsula.

With an air of detachment, beneath an innocent appearance as a charming man of the world, this new Gilles de Saintonge is a precise and demanding man of business — his talents no doubt sharpened by the ferocity of the luxury market in which he labored for several decades — meticulous, verifying everything, in a hurry, demanding, imperious about safety at sea and the assignments given to him by scientific institutions. By nature, and by his professional work experience, Gilles is media-savvy, a man so persuasive he can move mountains and magically open doors that would normally be closed to him. A modern adventurer, but above all a confirmed sailor, and therefore cautious, he is capable of safely conducting a maritime mission through the treacherous waters of the Caribbean. A passionate disciple, thanks to chance encounters and enthusiasms, he set to work for scientists seeking knowledge of the world through the refracting lens of the ocean. Measure the displacement of the Caribbean plate? Why not? Record the cries and whispers of tarpons in order to add to ichthyology's dictionary of dialects? Agreed. Because he has the gift for narrative suspense of Jules Verne, he can tell this story with great skill. His wonderment at the stars in the tropical sky during a night watch recalls a statement by Pedro Nunes, a Portuguese pilot of the first half of the sixteenth century: "We have discovered different islands, different lands, different seas, different peoples, and even more, a different sky and different stars." Because he seems to succeed at everything, Gilles Fonteneau is among those who have been lucky enough to realize their dreams.

An old and universal proverb reminds us that we get the luck we deserve.

Admiral François Bellec,
French Naval Academy

Prologue

The Venezuelan Army Plays *Waterworld*

"Take a look through the binoculars. Something's going on over there."

Two soldiers carrying Kalashnikovs were cautiously going down the long iron staircase toward the water, holding for dear life onto the platform's rickety handrail. Its fragile, rust-covered skeleton could be seen dimly through the spray.

Waves covered the lower steps and threatened to capsize the rubber dinghy that the Venezuelan soldiers had launched, unknown to us, while we were in the wardroom eating the excellent meal prepared by our captain, my namesake Gilles, and his sturdy son Joel.

The situation had suddenly become critical. This armed advance must have been the official response to our negotiations that had already been dragging on for three days and three nights.

Despite our countless official documents and our best efforts to reach an agreement, constant trips back and forth between the boat and this fortress that resembled a gigantic spider with twenty reinforced concrete legs had come to naught. We now had to make a decision . . . But it had all started out so well . . .

"Bienvenidos a la basa militar Simon Bolivar de Las Aves."

This had been our first contact, by radio, with the occupants of the military base, when we were only a few miles away, aboard the *Betelgeuse*, our converted tuna-fishing boat filled with scientific measuring devices. We were confident and proud of the work that we were to carry out with the researchers from two American universities traveling with us.

Waves caused by a strong easterly had given us a troubled night after we sailed from Guadeloupe. At dawn, everyone was on the bridge, eyes wide open and hair tousled by the wet wind. To the west lay a little strip of land bordered by a pale yellow sandy beach swept by the surf.

Everything would have been perfect had it not been for that damned radio contact. We thought no one had been there for years until we saw the steel-and-iron monstrosity emerging on the horizon, its form clashing with the flat, empty shore of the island of Las Aves.

The year before, in order to make sure that the island still existed, we had hired a small airplane to fly over it. While we were returning from this dangerous flight, the limited capacity of the fuel tanks had forced our pilot to glide to a landing with the engine cut.

Although its coordinates were uncertain, the two portable GPS devices we'd brought along helped us find the island without too much difficulty.

Flying over this strip of earth, we had in fact seen the metal superstructures. But we had seen no one moving on them and no one on the beaches. Only a few birds seemed to inhabit the ocean's solitude. We were proud of having proved that the island was still there, contrary to some pessimistic specialists who swore it had disappeared beneath the sea. We were also surprised to observe that it had not been split in two by the waves.

We were reassured: the research on the velocity at which the Caribbean tectonic plate was moving would indeed take place as planned. Our work might well provide information indispensable for predicting underwater earthquakes, and possible tsunamis, throughout the region.

We were now barely half a cable length from our island, battered by the wind. To port, we could see a long tongue of sand of about three hundred meters, with two little mounds of dead coral at the northern and southern tips. At its widest, the island must have measured fifty meters, at its narrowest thirty. We were anchored in a spot sheltered from the waves, in the middle of a cove on the island's west side. All the colors displayed by this strip of land surrounded by the ocean enchanted us; indigo mixed with the subtlest shades of emerald.

But to starboard, the piles topped with rusted steel cages were daily becoming more of a nightmare: not only because of their appearance, but also because of the band of nine soldiers living there.

We often had the impression that the structure came straight out of the movie *Waterworld*. On the platform, as rusted as the rest of the structure, several dozen containers were scattered around. They housed the soldiers of the Venezuelan army, for the island has belonged to the Republic of Venezuela since 1845. Topping everything, a small platform that must have been used as a helipad was connected to the main structure by a narrow staircase. A very old dock, probably leading to the island's largest beach, had been destroyed by the wind, so that the only approach to the platform was from the open sea.

It was March 20, 2002, the weather was very fine, and the air held a faint aroma of exotic flowers, despite the distance from the Caribbean islands. And we were about to be shot at.

MURMURS FROM THE DEEP

1

Las Aves: The Isle of Birds Is Sinking

The history of Las Aves — The discovery of a wreck — The mouth-to-mouth dance of the fish — Careful scientific preparation — The importance of the research — The television crew — Father Labat, a history linked with the history of La Rochelle — An island that will disappear — Political battles — The last roll of the dice.

Paris, a few months earlier

It was raining, and the cold gloom of the bustling city made me eager to leave as soon as possible. The bare entryway and reception room of the Venezuelan embassy offered me no comfort. The ambassador was away, and I had an appointment with the chargé d'affaires, who greeted me warmly. He had not only read of our expedition in the Venezuelan press, but had also encouraged us in our work, which he thought would be very useful for the Venezuelan authorities, his country's scientific community, and the preservation of the island itself.

Shaking my hand as he presented me with an official authorization, he wished me good luck: "You know, that island cannot disappear, or should not, but if it must, please make sure that it will be two hundred years from now!" He spoke to us as though we were in command and could do something about it. Things were certainly getting more mysterious as we went on.

While I was in Paris, I also delivered a cordial message to M. Cheminé, head of the Institut de Physique du Globe, from Professor Dixon, chairman of the geology and geophysics department of the school of atmospheric sciences at the University of Miami. We had put together our research program with him. A man of great kindness and simplicity, M. Cheminé was already aware of our work. He wished us good luck and gave me a research document: the geological cross section of the Caribbean tectonic plate, based on the work by the Centre Nationale de la Recherche Scientifique (CNRS), an expedition led by M. Mauffret and Mme. Leroy, researchers in the geotectonics department. He urged me to make contact with François, director of the observatory at La Soufrière de Saint-Claude in Guadeloupe.

The following day, I began research in the Bibliothèque Nationale, looking for historical information on the island of Las Aves. In a collection of maps from 1635 was one of the Caribbean on which the island appeared for the first time. Other more detailed maps confused our island with one much closer to the Venezuelan coast. Ours was clearly the one located 400 miles north of Venezuela and 110 miles west of Guadeloupe. But I was unable to determine who had discovered it. It was only a few months later that I came across additional information. At the time, this little scrap of land was obviously of little interest to explorers. Things have certainly changed since then.

Ponce de León might have sighted it in 1499. So might the Spanish navigator Alonso de Hojeda, who had commanded one of the ships on Columbus's second voyage, when he sailed in the area with the cosmographer Amerigo Vespucci. Or perhaps it was Jean, Sieur de la Haye.

Although we don't know who discovered it, later events are clearer. The first account comes from 1705, by Father Labat, a French priest who landed on the island after a shipwreck. He determined its geographic location, "fifty leagues to the leeward of Dominica and to the west of the great savanna," and continued that "what I can say about it is that this island is very beautiful, consisting almost entirely of sand and bushes, with very few trees. It might very well be called the Isle of Birds, for there are so many of them that they can be killed with sticks. There are also a large number of turtles, particularly when they lay their eggs. However, since the island is completely lacking in freshwater, no one comes there except by accident."

Then in 1771, another Frenchman, the Chevalier de Borda, determined its exact position by means of the chronometers on board the vessel *La Flore*, commanded by Verdun de la Crenne under the king's orders. He situated it "at forty leagues south southwest of Saba, and it is two or three leagues in circumference. Its elevation is eight *toises*, there are two neighboring rocks at a distance of a quarter league, no water but turtles and shrubs, guavas and sour sops."

With this established, in 1777 a Spanish decree gave

ownership to the governor-general of Venezuela, which made the island part of its national territory after independence. In 1978, Las Aves was declared a nature preserve, and a military base was built on it soon thereafter, which brings us back to our expedition.

Our aims were both simple and complicated. Simple because we intended to measure all conceivable dimensions of the island, and complicated because we had to install several satellite dishes to capture the signals that would enable us to measure the exact velocity of displacement of the Caribbean tectonic plate.

Half Sherlock Holmes, half Flash Gordon. What was all this work for, what good would it do? First, we are certain that the island will disappear before the end of the century. Consider the measurements that have been made by researchers since 1855. At that time it was 925 meters long; in 1939, 750 meters; in 1969, 530 meters; and in 2002, 300 meters. This erosion was the effect, of course, of wind, waves, storms, and changes in sea level, but not only those forces. That would be too simple. We have to go further, or rather deeper, to get more information.

The Caribbean plate has its own history. Four hundred million years ago, it was located in the middle of the Pacific Ocean. Sixty million years ago, it slid between North and South America, which were then separated. Since then it has calmly drifted, coming up against the Atlantic plate, and through a process of subduction or overlapping of plates, by folding the Atlantic plate it created the West Indies chain of volcanoes.

The faster the plate moves, the greater the risk of volcanic eruptions, because of the friction occurring at a depth of approximately 120 kilometers. By dissolving the earth's mantle, this friction allows magma to rise. And that explains our presence, to carry out that work, if only the Venezuelan soldiers would be kind enough to let us get on with it.

A final piece of information is necessary. Thirty million years ago, the angle of contact between the plates changed; the area of subduction was then located 110 miles west of the current area, just beneath our island. This cone, or rather magma chamber, not supplied with magma since that era, has subsequently had a clear tendency to shrink, collapse, and sink. What else can the island do, since it is immediately above the chamber?

Continental-drift theorists will criticize me for having oversimplified matters, but I am merely attempting to make things clear for a lay audience.

La Rochelle, October 2001

At the edge of the terrace of the aquarium restaurant, I surveyed all the boats moored in the trawlers' harbor, their straight masts like an army of lancers ready to leave at a moment's notice. Where did they come from? Where would they go? It was a fine autumn morning, a little cool, with the bluest of pure blue skies. To my right, the two towers, like old and tired sentinels, still stood erect as though to affirm their attachment to the city. It was just enough to make me forget why I was there.

One of the boats had already attracted my attention. An

old Breton tuna-fishing boat, with a white hull bordered by two blue stripes, of a goodly size and a solid appearance, was about twenty meters long and probably weighed ninety tons. That was just the boat I needed.

I had recently returned from the first year of my West Indies expedition, and having recognized the limitations of my first boat, the *Prince de Vendée*, a thirteen-meter catamaran, I was in search of another, more spacious one, because experience had taught me that I needed to take along more researchers and more equipment for our mission on the island of Las Aves.

The next day, I had a chance meeting with the owner of the boat, a very friendly man with the same first name as me, an experienced former fisherman, a solid and quiet-seeming man. He was preparing to leave for the West Indies in January, thence to sail around the world. His boat would be in Guadeloupe in March. He would wait for us there. We reached an agreement to rent the boat for a month; the equipment was excellent, even including a compressor to recharge our diving tanks. So, we had a new boat, *Betelgeuse*, named after the red star marking the eastern shoulder of the constellation Orion, very visible to the south in the Northern Hemisphere.

It remained to organize the arrival of the researchers at the right date at the airport in Pointe-à-Pitre, to inform them of the space available on board and the possibilities for loading scientific equipment, and then to alert the Paris press and our friend Claude, of the *Sud-Ouest* newspaper, of the date of our departure.

Finally, we had to provide our sponsor, the Bacardi Family Foundation, with all the information necessary for monitoring our mission on the ground. We also had the pleasant surprise of seeing a good article in *Le Figaro* describing our expedition. The title was a perfect capsule of the situation: "In the Caribbean, the Isle of Birds Is Sinking."

Through my friend Yves, a journalist at *Ouest-France*, I was surprised to learn that a television crew from the French Overseas Network in Guadeloupe wanted to hook up with us, to prepare a one-hour documentary. I also contacted a friend named George to be our photographer on board.

We were scheduled to cast off from Guadeloupe on March 16, 2002. I planned to arrive ten days in advance. We had been working on this project for two years now, and it finally looked as if it might succeed.

After loading two bags of diving material onto the *Betelgeuse* that would cross the Atlantic without me, I prepared for departure, putting together a detailed work schedule for each of us. The work consisted of installing two receiving antennas for each satellite, each one connected by a computer; measuring the island in all dimensions; studying the corals in the greatest possible detail; recording by hydrophone the language of a few species of fish; trying to situate the wreck described by Father Labat; listing the flora and fauna; verifying depth measurements; and finally, cleaning up the shore and recording the origin of all the debris that we might find: enough to keep us busy around the clock.

* * *

Guadeloupe, March 9, 2002

As always, as we emerged from the plane we were greeted by an oppressive, humid atmosphere filled with the aroma of very sweet ripe fruit. The croaking of little frogs was more insistent than ever. There was no question that we were in the Caribbean. There was a hint of adventure in the air, and we could only hope that it promised good fortune. What were we finally going to discover on the isle of Las Aves that everyone wanted to keep mysteriously secret?

Eric, the television producer, was waiting for us and drove us immediately to the little hotel where we had made reservations near the marina of Bas du Fort. I told him about our research program, the dates for the boat's sailing, the space available on board for his crew, the official authorization from the embassy, and all the possible details that might help him in his work.

He had already planned to interview me that evening on the local television news program. He described his plans for the film, and everything seemed to be getting off on the right foot. I sent e-mail confirmations to the crew of the dates and times for our sailing. I called our captain on his mobile phone to make sure that he had made it safely across the Atlantic. The *Betelgeuse* had put into port opposite the Îles des Saintes, and would be at the harbormaster's office in the marina of Bas du Fort on the appointed day to take us and all our equipment on board.

Two days later, I confirmed the exact date of our arrival at the island to Christian, the pilot of the Hughes 500 helicopter who was supposed to take aerial photographs of Las Aves. He would then work out his flight plan and ask the aviation authorities in Puerto Rico for authorization. He also told us that because of the great distance between Guadeloupe and Las Aves, we would have to carry two hundred liters of kerosene on the boat so that he could refill his tanks when he landed on the island.

I had received several messages from Mr. O'Brien, president of the Bacardi Family Foundation in Washington, who intended to celebrate our sailing with a cocktail party on the dock for local employees and customers of the company. Of course, he assured me, a television crew would be there.

My fifteen years' business experience in North America, as head of operations in the Western Hemisphere for a major French perfume company, stood me in good stead in making sure there were no slipups.

We also had to organize the delivery of cement, gravel, crowbars, wooden panels, and batteries, find a supermarket near the dock so that everything would be ready on time for our departure, and provide enough food for ten people for ten days.

There were also e-mails to answer from the University of Caracas, which was supposed to send us a specialist in tectonic plates. Because of insufficient resources, and to my great regret, this plan was abandoned at the last minute. Then I had to plan my flight back to Miami to meet Professor Tim Dixon, the originator of the program. I had to plan my future activities in the islands to present the lectures I was supposed to give on Saint Martin, and answer the marine nature preserve on Saint Barts, which had asked me to continue my research on the sounds made by groupers, using its boats. And of course, I had to make sure that nothing was missing.

Time went by very quickly, and I was on board with everyone on the eve of sailing, prepared to smile and explain the purpose of our expedition to anyone who wanted to listen at the cocktail party organized by the Bacardi Family Foundation. The television crew had already stowed two hundred kilos of equipment on board; the sound engineer and the cameraman were very cordial. Everyone had located his bunk, I had an adequate supply of film, all my equipment was on board, and we were finally ready.

The evening was a great success. Mr. O'Brien, delighted by the media presence, asked me to start working with his organization at the end of the year to organize a plan for a nature preserve in the Bahamas.

But I was thinking about other things just then. The success of this plan, which had taken more than two years of preparation, was an absolute priority, and I was constantly wondering what the island had to hide from us, because I had received so much contradictory information about it.

The next day, we cast off at the appointed time. I wondered as we sailed whether everything I had done up to then came from my search for adventure, or whether my taste for adventure came from the pursuit of research. Probably a mixture of the two, I suspect.

Our crew consisted of two Americans, two men from a village near La Rochelle, one from Martinique, one from Guadeloupe, one from Lorraine, one from Paris, and two from the Vendée. We had set all the GPS instruments on board at 63°36'49" west longitude and 15°40'30" north latitude. That was where everything would begin.

We left the little Isle of Pigs to starboard around noon. We then turned into the wind to raise the mainsail and the mizzen and, turning eighty degrees to starboard, unfurled the large jib. There was no winch, so we had to do everything with the strength of our arms. Our boat picked up speed under a clear blue sky, and a slight swell from the west did nothing to interfere with the cameras.

To port, we could easily see Marie-Galante, a small, completely round island, where a few years earlier I had had the good fortune to meet an extraordinarily kind and wise man. Infected with leprosy at the age of eighteen, he had essentially been rejected by his fiancée's family in Basse-Terre. He had decided to seek treatment in France, in Toulouse, if memory serves, and later returned to buy some very rocky land, which he had completely cleared and transformed into a sugarcane field that provided him with a modest income for the rest of his life. I wondered what had become of him since then. Was his rum better than the others? Was it stronger or sweeter than the rum that Father Labat had drunk? At the time, I had found a Creole title for an article: "Les Îles qui loin ka charmé moin" (The distant islands that have charmed me).

To starboard, as we sailed past the nature preserve of Basse-Terre, the clear sky allowed us to admire the deep green of the pass of Mammelles, the peak of Bouillante, and above all, La Soufrière.

This volcano towers 1,467 meters over the Caribbean. It sleeps with one eye closed, and winks with the other to warn at every moment of its possible awakening. Would our work provide the few bits of information enabling more accurate prediction of future eruptions? A few days earlier we had gone to see its dome and its sulfur springs. This was our way of communicating with the volcano, seeking its goodwill. We certainly would have liked to predict its next eruption, but others who were much better equipped and more capable had it under constant watch from the observatory of Saint-Claude.

From the Saintes canal, Las Aves was straight ahead. While sailing at a rapid clip, we put our dinghy in the water to enable the television crew to film the boat under full sail. All we had to do was let the wind carry us.

At that point, everyone could organize his life on board — some cooked dinner, others checked their equipment. For my part, I went over in detail our coming work schedule with the team of scientists. The following wind made our boat very stable, in spite of the swell that had already been present for several hours in the channel between Guadeloupe and the Saintes. The *Betelgeuse* was plowing its white wake beneath the starry sky.

The next morning, everyone was on the bridge; Las Aves was in sight. The metal superstructures came into view first on the western horizon; even though navigating to that minuscule spot was fairly easy, we were proud that we could finally see it. We had been told before leaving that many boats had tried to find it and failed. Perhaps their equipment wasn't adequate, or their captains not as good as ours.

In any event, in a few hours we would have the answers to many questions that I had been asking for the last two years.

Just a few miles from our destination, the onboard radio came alive. A barely audible voice welcomed us in Spanish. So the island was not deserted? Who was on it? And what were they doing on this remote island? I put on my best Spanish accent to thank them for their radio welcome, and to inform them of our mission and our intention to drop anchor in their waters.

Two or three hours later, we came in view of that huge spider with cement legs towering over the unspoiled island. Our interlocutor recommended that we moor in the cove to the west, leeward, and half a cable from the monster. We dropped anchor at a depth of ten meters. The sea was calm; spread out before us was a beautiful crescent beach of pale yellow sand. At the northern and southern ends, two mounds covered with a mixture of little pieces of dead coral and guano rose to a height no greater than three meters above sea level. A stretch of sand barely one meter high separated the two extremities, and its faded appearance clearly showed that the sea had certainly covered that spot in the past. A few cormorants and frigate birds whirled across the cloudless sky. Complete silence prevailed. On board, I was already mapping out a plan for installing our various devices. As for the cameramen, they were itching to start filming the operation.

We were invited to come to the fortress to introduce ourselves. When we got to the platform, eight soldiers and their commanding officer were standing at attention, saluting us in a very official way. They were all wearing off-white uniforms, a kind of work clothes, and they all seemed to be in their twenties. We must have been an unusual sight for them, something that would for a few days relieve the monotony these young men must have experienced on the isolated island.

After the touching ceremony, the commanding officer invited us to tour the base. We stood high over the island. From the railing, worn away by rust, I could see both sides of the island; that's how narrow it was. I was able to verify that the southern end had been completely washed away by the sea, already showing huge long strips of stratified limestone. Just below us, two poles that must have been used for a volleyball court stood amid visible tracks of the sea turtles that usually came to lay their eggs on this beach. The central, even narrower, part of the island had no trace of vegetation. The sea must have covered the area during severe storms, dividing our island for the moment in two.

To the north, a small coral reef protected a tiny beach. Because of the short distance, I could clearly see the earlier mentioned mound of dead coral and guano that was probably used for nesting by the birds. The whole eastern coast was battered by the sea, which had swept away a kind of coral reef that I had seen from the air the year before. I no longer had doubts about the future disappearance of Las Aves. Our promised land seemed very fragile in the midst of that pitiless ocean. Our evidence might therefore be useful in 2002, either to preserve it or to demonstrate its slow and certain disappearance.

We were soon asked to join the commanding officer, who wanted to meet with us inside one of the containers used as a meeting room. A stale odor of wax and dampness, mixed with the smell of fried food, greeted us. To the right of the entry door, which separated it from a long narrow kitchen, was a television set. The walls were lined with dark plywood on which were hung a number of photographs showing a visit by President Chavez sometime earlier. A framed, coarsely printed document, which we saw clearly only when we got used to the darkness, provided a brief history of the island. To the left and toward the back was a large rustic table surrounded by a dozen plastic office chairs with stainless steel frames.

El Commandante offered us coffee, while another soldier was attentively watching the television set. Glancing at it, I got the impression that there was a revolution somewhere in South America — nothing out of the ordinary. Official introductions could now begin. Each of us could describe his duties. I explained as simply as I could the work that we planned to do, under his strict supervision, of course, and following the instructions and recommendations that he would be kind enough to give us.

He had to be reassured, for his air of a nervous peasant soldier left no doubt as to his scientific ignorance. Our conversation, in Spanish, was simultaneously courteous and suspicious. He relaxed a little when I gave him the letter from his embassy in Paris.

During our conversation, interrupted by unexplained silences, we had very mixed feelings. We had no idea in what direction things were moving with him. He seemed to have complete control over the situation, but was unable to make any decision, which did not augur well for our prospects.

After two hours of discussion, interspersed with glances

at the television set, the commander finally informed us that he had to follow procedures and inform his superiors on the mainland. If the radio was in operating order, he would give us a decision as soon as possible. Meanwhile, we were free to remain at anchor, but we should not go too far from the boat. We could put our dinghy in the water, but under no circumstances were we to land on the island without his authorization. He needed time, and he would contact us by radio. As a parting gift, we presented him and his crew with a case of good Bacardi rum.

On the way back to the *Betelgeuse*, the conversation was animated. We had to modify our program. Part of the crew decided to explore the underwater northern tip of the island, and some of them took fishing equipment in the hope of improving the evening's dinner. I asked them to take the opportunity to survey the condition of the coral and take note of the species of fish they found there.

Another crew member and I decided to use diving tanks to take a look at the depths on either side of the boat. The water was extremely clear. From the surface, we could already see that the bottom was almost empty. Diving into the silence, broken only by the noise of the bubbles coming out of our regulators, among the few triggerfish and parrotfish going about their business, we soon realized that we were surveying an area in which the entire bottom had been thoroughly stirred up by the sea. We were in fact in the area extending from the narrowest part of the island.

We could thus conclude that the ocean, which periodically swept over that part of the island, had also in some way washed away the bottom. This indicates the force of some hurricanes, like Luis, for example, which struck in September 1996. This also explained the carpet of dead coral covering the bottom. Here and there a few signs of life still survived. Skimming over the bottom and staring attentively, by pure chance I found a little block, fifteen centimeters on a side, barely buried in the sand. When I lifted it, I realized that it was a piece of brick covered in limestone. The rectangular shape left no doubt as to its origin; it could not be a product of undersea patterns of life. Could I have put my hands on a vestige of what happened in January 1705?

On January 8 of that year, following a strong storm that had driven him into the area, Father Labat found himself in sight of our island. "A leeward land belonging to us," he wrote in his account, *Voyage aux isles*. This bold priest, who left La Rochelle in late 1693 to preach the gospel in the Leeward Islands, and to whom we owe a unique description of the customs of the time, was indeed in the neighborhood of Las Aves one fine day in January 1705. He describes in detail what he encountered there: "On the eastern shore we saw a ship lying on its side and ten or twelve men on the sand who seemed to be English. By their gestures they seemed to be asking us for help. These unfortunates had been there for eleven days. There were fourteen men, with two respectable women, and eight slaves, male and female. They had run aground because they had not been aware of the island."

The vessel came from England and had put in at Barbados. Father Labat goes on to tell us that the captain had taken a launch to look for help, that everyone had removed the contents of the hold to the island after discovering that the hull had been breached in two places.

The boat was therefore lost. The recovery of everything that might be useful lasted for ten days, in the course of which the masts, sails, cannons, anchors, tin and lead ballast, cloth, salted meat, Madeira, ropes, canvas, cases of hats, and everything that would fit was loaded onto the boat on which the good father was sailing. While this was going on, he learned that some of them had buried chests containing precious merchandise and silver in one corner of the island. The secret was betrayed and the chests were found and put on board.

Finally, after negotiating conditions, everyone embarked and landed a few days later on Saint Kitts and Nevis. The completely empty hull lying on its side remained, and since then, particularly because it rested on the windward side, it must have altogether disappeared, broken apart by bad weather. It had happened nearly three centuries earlier, and we therefore had no hope of finding any traces.

But the unexpected happened that March day of 2002. I soon realized as I examined my find back on board that it was a piece of brick used at the time to frame the cooking hearth located on the bridge. The limestone deposit covering it was about three centuries old, which was confirmed by later analysis in France. The brick had thus been displaced by a good hundred meters since the wreck, moving from the windward side to the spot where we had found it. Having discovered something that reminded us of that adventurous time, we found Father Labat even more attractive. In the course of the evening, all our thoughts focused on him. What more would we discover? The next day we would have to explore the area more thoroughly and dive on the east side of our island.

The onboard radio immediately put an end to our discussion; El Commandante wanted to see us all again the next day at dawn, and we had to bring along our passports. After dinner, in the course of which everyone made suggestions for improving our prospects and for which attitude to adopt, I decided to listen to the news on the shortwave service of Radio France Internationale. The day before, there had been a coup d'état in Caracas, the capital of Venezuela. President Chavez had taken flight, the population was in the streets, the army had fired shots, and there were a number of dead and wounded. This disturbing news made us wonder how it would affect our expedition.

When we reached the top of the stairway, through whose holes, eaten away by the sea, we could see the emerald green of the ocean, we were again taken directly to the meeting room. We had decided to keep this meeting small. In the room, three soldiers were standing in front of the television set. Their commander was absent, apparently communicating with the Venezuelan authorities, and thus enabled us to confirm the accuracy of the news we had heard on television. The broadcast showed there was widespread panic in the country, and the commentator himself seemed unsure what was going on, as the crowds alternated between joy and hostility. Not good news for us, I feared. El Commandante entered and greeted us, but made no comment on the current situation. He had a blank sheet of paper and asked us if we all had our passports. He seemed preoccupied but tried not to show it. After noting down for an hour all the information he needed about our respective identities, he informed us that he intended to keep our passports until the following day. He also informed us of the difficulty he was experiencing reaching his superiors, because his radiotelephone was not working, and told us that, therefore, he could make no decisions.

I argued that we had important scientific work to do, that his country would be the first to benefit from it, that we had come from a great distance, that it had taken two years to prepare the expedition, that his embassy was impatiently awaiting the results, that we had no weapons on board, that we were all concerned with the preservation of the island, that our work was purely scientific with the purpose of predicting earthquakes in the region, that the University of Caracas was closely following our operation, that various other universities around the world were awaiting our results, that we needed only ten days, after which we would leave, that he himself would be rewarded by his government if he allowed us to conduct our research, and God knows what else — every logical argument I could find to persuade him.

Although he listened to me attentively, I sensed that it was all way over his head. Orders are orders, and without approval from his superiors we had to keep on waiting. He reiterated that we could not land on the island and that the situation remained the same as it had been the day before. After that, he cut the conversation short, said he would contact us through the onboard radio, and exited the room, leaving us with the same few soldiers glued to their television set. They cordially offered us coffee, then took us to their favorite fishing spot, the platform railing just below the helipad, where they fished every morning.

We established an immediate rapport with them; friendly simple soldiers with no great fondness for their military service. They told us that their commanding officer had just been appointed to the post and that his lack of experience was also causing them some problems. Because of his interest in science, their former commander, who had just left, would have let us do our work without difficulty. They also told us that, if allowed, they would be happy and proud to help us.

Meanwhile, the television crew decided to book the dinghy and take some exterior shots. The captain and I lugged equipment for underwater photographs of the northern slope the divers had noted the day before. After going down a gentle slope decorated with tubular sponges, anemones, sea fans, and brain coral, we reached a depth of twenty-five meters. This world was populated by varieties of fish — butterfly fish, sabre squirrel fish, blue-headed rainbow wrasse, angelfish, and finally a multitude of chromis.

Against the blue background, a huge school of shimmering common jacks swam around us. At any moment, one of them might give the alarm and, in a swift movement, the school could disappear in a flash. A few lobsters, startled in their slow, deliberate progress, quickly disappeared with a flick of the tail into their dens. At one point, two bluestriped grunts, each one weighing about a kilo, posed in front of my camera as though they were particularly proud of their makeup and wanted to pass it on to posterity.

Then a strange ballet was performed before my eyes. The two fish each moved away from each other for a moment, then turned and came face to face, mouth to mouth! Moving apart again, they repeated this display three or four times. Fortunately my camera was able to make a permanent record of this scene, which I found incomprehensible. What did this posturing mean, and what were the reasons for it? I was both moved and excited that I had witnessed such a spectacle. They resumed moving quietly side by side, seeming to watch each other from a close distance, and then they repeated the same dance. I was then able to see their bodies trembling when they came face to face. I took one shot after another. I hoped that the light was good and the pictures would come out.

It was only a month later, when I showed the pictures to Professor Myberg, a marine biologist at the University of Miami, that I got the answers to my questions. I had in fact heard about this behavior but had never seen pictures of it. These were two males, *Haemulon sciurus*, who, by means of quiet but serious aggression, wanted to mark off their territory; all of this was no doubt connected with the females inhabiting the area. It should also be said that these bluestriped grunts emit sounds that we had recorded the year before in the lagoons of the island of Barbuda. My photos were published in the July 2002 French edition of *National Geographic.*

We used the afternoon to go over our scientific equipment and unsuccessfully to try to contact the University of Caracas on our satellite phone. At six in the evening, the base called us to set our next meeting for early the coming morning.

The soldiers greeted us at the top step of the stairway. The air was crisp that morning, and their faces were impenetrable. Something new had probably happened. There was the same aroma of wax, the same mismatched furniture, the same decorations on the walls; I then noticed a rather faded picture of the Virgin Mary near which were hung a few plastic roses, which had earlier escaped my attention. The commanding officer again kept us waiting. When he sat down with us fifteen minutes later, I noticed his mannerisms. He had an uncontrolled habit of moving his chin. He reminded me of the final guardian of the western frontier sitting at his desk, half epileptic and half mad, whom I had seen in Kevin Costner's film *Dances with Wolves*.

Everyone was on edge. My first concern was to recover our passports. In the ensuing silence, barely troubled by the television still giving confusing news about the situation in the capital, the commander's gaze became increasingly abstracted. He hesitated to speak while he was watching the TV. Finally he informed us he had still been unable to reach his superiors. Therefore, we had until the following morning to raise anchor and leave the island. This was clearly an ultimatum. I tried to argue for just two more days, which would enable him to make contact, but he wouldn't hear of it: we had to leave as quickly as possible, he reiterated.

Back on board, and feeling thoroughly helpless, we decided to hold a council. Did this incomprehensible decision come from him, or had he hidden its source from us? In any event, in the next few hours, thanks to our satellite communications system, we contacted all the media outlets with which we had relationships. I gave a live radio interview on our current situation, expressing my failure to understand the decisions made by the base. We also contacted Agence France-Presse, so that it could quickly spread worldwide the facts of our bizarre situation. Then we called the pilot of our helicopter to inform him that he should stand by for further information from us before joining us. The television crew filmed my protestations, which I still vividly remember.

After a good deal of hesitation, we decided not to give in to their ultimatum, to allow time to work in our favor. Within twenty-four hours we would know the outcome of the duel that was about to begin. We were all both angry and disappointed. I was probably more shaken than the others. All that work, all that preparation, all that organization, and all those resources now seemed to be part of a nightmare.

The next morning at nine, when our ultimatum expired, nothing happened. The onboard radio was silent and everyone sat tight. Over the course of the next few hours, we experienced a mixture of hope and despair, until the moment when our captain told us to look through the binoculars at the two armed soldiers who were about to board their flimsy craft. In order to avoid the worst, we decided with sinking hearts to lift anchor. Having done nothing, we would leave the soldiers to play volleyball on the tortoise nests. Our return to Guadeloupe was extremely difficult. Not only were we sailing into the wind, which was creating a large swell, but we were also bearing the sad fact of our failure.

François, director of the Soufrière observatory, greeted us cordially and told us how sorry he was for our problems. He also told us that a few years earlier the French government had asked the Venezuelan government for authorization to install a seismic measuring station on Las Aves, and that the Venezuelans had refused, adding that between March 2001 and March 2002, there had been 549 tectonic tremors between Martinique and Saint Martin, forty-eight of them on La Soufrière, ten of which had been perceptible. Some of them had taken place at a depth between fifty and one hundred kilometers.

Specialists on the region informed us that the ultimatum given to us was perhaps not at all legitimate. In fact, the governments of Dominica and Venezuela, both of which have good reasons to claim sovereignty over this isolated patch of sand in the middle of the ocean, had not yet resolved their dispute. The countries of the eastern Caribbean asserted that Las Aves did not really satisfy the criteria for an island territory. The claim by the Venezuelan government encroached on the exclusive economic zone of the countries of the West Indies, which meant that one day Venezuela might claim authority over the maritime territories lying between Grenada and Montserrat, in particular on fishing areas. At the present time, that area covers more than 150,000 square kilometers.

The government of Antigua has strongly asserted that acceptance of Venezuela's claim would be completely contrary to the United Nations convention on the law of the sea. Dominica is said to be prepared to demonstrate the existence of geological links between it and Las Aves. Aside from the fact that the island represents a huge fishing area, we also learned that the Venezuelan government hoped to find oil and gas there.

President Chavez had already made known his answer to all these arguments by neighboring countries a year earlier: "No one can deny Venezuelan ownership of the island of Las Aves. In fact, we are going to increase our military presence there in order to expand our rights in the territorial waters and the exclusive economic zone." The heir of Bolivar has retained the lesson of his master: "If nature opposes our aims, we will fight to make it obey us."

Someday this dispute will doubtless come before an international court. And yet France had come close to appropriating the territory. The French government, in the boundary treaty signed on July 17, 1982, by Olivier Stirn on behalf of the Foreign Ministry and Gustavo Planchet Marique representing the Venezuelan government, ceded the island to Venezuela. This bargain, made without any counterpart, to our knowledge, was incomprehensible. In 1993, a fishing boat had been mysteriously bombed by a plane north of the island. The transfer now poses a problem about the western maritime boundary of the Lesser Antilles chain. In fact, measuring two hundred miles from Las Aves and two hundred miles from its own coast, Venezuela now claims jurisdiction over a territory extending four hundred miles to the north of its coast, which is challenged by all the countries adjoining the island. Between this legal tangle and the coup d'état, our chances in the year 2002 had been seriously damaged. The risk of letting us work had nothing to do with the ongoing arguments, except if we had been able in the course of our expedition to determine more precisely the rate at which Las Aves was disappearing.

When will we have the opportunity to return in better circumstances? What will be left of this scrap of land? What will happen to the spider with feet of cement? And will its commanding officer one day regret his decision, or had he lied to us from the very beginning? These were the questions running through my mind as the plane took off for Miami.

We were all very disappointed, and I hoped that my extraordinary crew would not hold it against me. Other adventures were now being considered.

As for *Betelgeuse* and my namesake Captain Gilles, they continued their voyage around the world, setting the course for the ship's white prow and its Oregon pine bowsprit through the Panama Canal, across the Pacific, through the Torres Strait, and the Red Sea, finally back to its home port of La Rochelle in June 2003. I was there to welcome them. When the little white plume that the prow created as it plowed through the ocean turned into a ripple as it entered the channel leading to the two towers of the old port, I could finally see that Captain Gilles, having barely reached the end of his journey, still had his eyes filled with distant landscapes. Recalling our adventures, he told me that Las Aves was in the end one of his happy memories. For my part, every time I see a steel cage, a stale aroma of damp wax assaults my nose. I cannot drive out of my mind those two armed soldiers going down the long rusted stairway leading to the Caribbean Sea.

2

Listening to the Silent World

The underwater orchestra — The discovery of the language of fish — Night on the Intracoastal Waterway — NASA's boats — The launch of Atlantis *— Mad scientists — Equipment to begin with — Contacts with the University of Miami — A center for oceanographic research — Corals and tectonic plates — The first research programs — The sponsor is found — We pack everything up — A comfortable return — Crossing the North Atlantic — The first tests of portable GPS devices — The discovery of the* Prince de Vendée *— The electronic equipment on board — The installation of a laboratory at sea.*

Cape Canaveral, February 15, 2000

With a gentle plop the hydrophone slipped below the calm surface of the murky water, and the concert began. The whole team of researchers could hear it through the speaker set up on the bench at the stern of the flimsy NASA boat.

It was seven in the evening, and the blue Florida sky had

just lit up with the orange streaks of sunset — the spectacle could begin.

This novel score began with a repeated series of deep booms accompanied by dry staccato bursts. It was as though an army of double basses were playing in chorus against a background of drums being beaten with cotton-covered sticks.

There followed the contralto of a foghorn droning in a barely rhythmic pattern. Beneath what sounded cacophonous to me was a constant noise of crisp staccato hissing, like bacon sizzling in a pan. No one else seemed surprised, quite the contrary. With each different note, our friends nodded as they recognized a sound that seemed familiar to them. It was as though I were at a Wagner concert for the first time in my life. Just below, dozens of trumpet-fish were there a few guitar-fish. I suppose I belonged to a different world, the world of earth dwellers. The others seemed to have gifts that I did not; perhaps they understood the language of fish.

We were surrounded by a huge mangrove forest that covered the northern part of what was appropriately named Mosquito Bay, on one of the canals that ran between an inland waterway and the ocean. This spot is within the nature preserve belonging to Cape Canaveral. Clearly, these people, with their heads in the clouds, were also interested in what went on underwater.

This was my first experience of this unknown world, and I wanted to know more about it immediately. I had a lot to learn before I could make my own recordings.

"We are hearing at least five different species of fish and

other animals living in the bay," Professor Gilmore, a NASA ichthyologist, told me. "The female fish are now laying eggs, and what we hear are the males calling them for a mating dance, a ritual that always follows the same pattern. The sea trout, the silver perch, and the anglerfish, common species in Florida and the Gulf of Mexico, made the sounds that you heard; the frying pan was a bonus generously provided by some shrimp that also chose this spot to reproduce."

I was simply astonished. I imagined that one day perhaps, by listening to them, I would be able to interpret their sounds in the same way, as long as I was trained to do it. But I was far from that point.

We were in the midst of romance. The males indeed always choose nights when the moon is full for their serenade, allowing the females to release their eggs while they spread their seed. Everything then blended together in a reproductive cloud, the slightly opaque veil of a new cycle of life.

The professor went on: "You know, Gilles, you have to realize that the ocean is a very dark place, because the average depth is four thousand meters. Light reaches two hundred meters, sometimes four hundred, but no deeper. That is why fish use sound to communicate, because they can't see. We are too accustomed to light to understand the world of the sea that depends much less on it. Fish are able to determine what is around them merely by analyzing sound frequencies."

How and when do fish use these sounds? How do the females react? What is the dynamic between predator and prey? These are the questions that his research team is working on. To do this, they have developed a computer program that, by separating each sound, makes it possible to calculate the intensity in decibels and the frequency level. I asked him, "Then why is NASA so interested in your work?"

"NASA doesn't focus only on space. We have many scientific programs. If we don't understand how plants and animals live on earth and in the oceans, we'll never learn how life can adapt to space. If one day we have to travel to Mars, first we'll have to understand how we live on earth. We are also interested in all the new technologies that might enable us to use sound to penetrate the mystery of the glacial caps on Europa, Jupiter's moon."

I thought that this method, which made it possible to detect reproductive zones, could also guarantee that they were protected.

We had already recorded a number of sounds, and it was time to move on. After measuring salinity, temperature, and oxygenation, we submerged the microphone again. We now heard the creaking of a rusty old door closing, followed by a quieter dry clicking. "It's a dolphin," he said, "who has detected the sea trout from the sounds they're making."

I stayed with Professor Gilmore's team for a week, not only to familiarize myself with all the apparatus, but also to learn how to be an apprentice ethologist. I had the pleasure of spending time with his family, meeting other researchers, and trying out everything, so that once I was on my boat I would be able to send him reliable information, because he had agreed that our expedition could proudly carry NASA's colors. Very early the last morning, just before returning to Miami, I was ready to watch the launch of the space shuttle *Atlantis*. After a huge roar that made the earth shake, the rocket rose slowly into the atmosphere, cutting through the dry air with a harsh, tearing noise. The day before at lunch, a young researcher had matter-of-factly explained to me how he had calculated the weight of fuel needed to blast off. Listening to this otherworldly uproar, I wondered if the nearby fish might be thinking that men were mad.

On the road back to Miami, I had five hours to think about what had just happened to me, which reminded me of a trip I had made for my work two or three years earlier, south of the Monterey Peninsula in California. Friends had taken me along on one of their expeditions to the top of the cliffs overlooking the Pacific to bird-watch. In the distance, amid the seaweed, a few little sea otters were settled comfortably on their backs, using their teeth to break the shells of the mollusks for their meal. There were no birds in sight, and yet one of my friends, an occasional ornithologist, said to me, "You see, this year those crazy brown-footed gannets came earlier than expected." He went on, "The puffins are there, too, right down below."

I had seen nothing, and yet he had heard everything and determined that the birds were there, identifying them through their cry alone. There was something magical about it. Maybe in the not too distant future, we might have the same experience with fish.

The stay with Professor Gilmore enabled me to establish our method of communication once I was at sea. With the rest of the material, he would send me a large number of minidisks, which I would return to him with a document prepared by the two of us, indicating the exact geographic location of the recording, and the salinity and temperature of the water, along with, of course, the names of the species recorded. We were to concentrate solely on groupers, porgies, and swordfish.

I gave him our schedule, which would begin with a visit to Barbuda early the following year, where we would make our first recordings. For him to give me such a responsibility, I must have earned his trust. I arrived in Miami full of enthusiasm, with my third research program in my pocket. At the time, I could never have imagined the adventures that lay ahead.

The idea for this expedition was the result of a development in my professional life and the long-standing contact I had with the marine science school of the University of Miami.

I had decided to buy a thirteen-meter catamaran to sail quietly around the Caribbean islands for two years. The director of the science department, whom I had been able to help through my facility with languages, suggested that I add usefulness to pleasure by undertaking some scientific tasks on the trip. A meeting with researcher friends immediately confirmed his opinion. Thereafter, Professor Gunsberg, a world-renowned specialist in corals, took an interest in the plan and asked me to sit in on his classes for a few months. I hardly became a specialist, but I acquired enough knowledge to understand, for example, the mechanism by which corals reproduce, and to familiarize myself with the role of zooxanthellae in the synthesis of the material that enables the coral to grow one centimeter a year. This was how I secured my first research contract with the National Oceanographic and Atmospheric Administration (NOAA). With the help of documents supplied by them, I was to provide as precise a description as possible of the situation of the coral reefs in the Caribbean. I was also asked to verify that there had been a whitening of the surfaces of some coral reefs that had already been detected by satellite. I was too happy to ask them for money; everything would be done at no charge.

The second contract came about by pure chance. During my classes, in the most important marine biology school in the United States, I made the acquaintance of a student working in the department of geology and geophysics. In the course of a conversation about continental drift, I asked him if he had a precise idea of the position the continents would be in in a million years. Probably preferring to defer, he suggested I meet Professor Dixon, a world-renowned specialist on the subject. In fact, he added, he happened to know the professor was in his office, right upstairs. Dixon had just finished lunch and was having a cup of coffee. I introduced myself, told him why I had come, and described my planned expedition in the Caribbean. Busy at the time, he gave me an appointment for the following day.

When we met, *he* was the one asking the questions. He wanted to know everything about the size of the boat, the itinerary, the dates, and so on. His geophysics department had a proposal to set up several GPSs on the island of Dominica as well as on an isolated little island I had never heard of called Las Aves. These instruments are used by satellites to measure the movement of tectonic plates and the speed at which volcanoes swell. He told me that if I agreed, he would be glad to meet with me at greater length to get a clearer picture of the proposal. Once again, I was sitting in on classes and familiarizing myself with barbaric terms like subduction, magma, displacement speed, and hot spots. The second contract was signed, of course with the same conditions. However, while on board, the researchers would pay for their food. Many other details were discussed over the following weeks.

That changed everything. With these three research programs, I now had to work out in detail the most precise itinerary possible, thoroughly inspect the equipment on my boat, and find some additional subsidies to pay for all of it. While passionately interested in this new aspect of my adventure, I realized that I had to give serious thought about how to publicize the affair. I had been fortunate enough to gain the trust of many people, and I couldn't let any of them down. The public relations department of the university was already interested in my expedition and had talked about it to the local press.

In the meantime, I was on the phone to La Rochelle, where the boat was being built, checking on its progress. In view of the latest developments, I decided to increase the power of the diesel engine from 27 to 37 horsepower and to supplement the electric power with four solar panels. I also learned that there might be a delay in launching.

It was time to look for a sponsor. The Bacardi family had established its rum company in Cuba in 1862. In 1960, it had been expropriated by Castro. Since then, through courage, hard work, and ingenuity, it had once again become one of the largest private companies in the world. Throughout its history, the Bacardi Foundation had always fiercely devoted itself to protecting the oceans, having financed a number of research projects in the Bahamas.

It was fairly easy for me to make contact with Joaquim Bacardi, the young and brilliant international chairman of the company — a few key words like "Cousteau," "NASA," and "expedition" were enough to secure a meeting for the following day. Armed with the research program and letters of recommendation, I met Joaquim in his office with the president of the foundation, Robert O'Brien. After listening to me carefully, they asked me to prepare a budget so that they could consider their financial participation. This was done and accepted two weeks later. All the details of our relationship were settled. They were going to put me in touch with their public relations office in New York, share the cost of the electronic equipment on the boat, organize a series of lectures that we would give in the Caribbean for their agents, and provide a sum of money for which we would have to account every month. We thus had a sponsor, and the future showed that this was the best choice we could have made. This foundation was an invaluable support for us, not only by keeping all its promises, but also by constantly encouraging us in our work.

All that remained was for me to find the few scientific devices I needed, because I was the only one who would be listening to the fish. I had to settle a thousand details about our organization.

Locally, I found good equipment for underwater photography, all the navigation books and reference books for the research, all the devices needed for onboard cooking, bad weather equipment, and other indispensable items for an operation of this kind.

I decided that there would be two kinds of crew members on board: those who would cross the Atlantic with me and those I would recruit locally once I reached Antigua.

Because I was well acquainted, from past experience, with the islands and their moorings, I was easily able to draw up a detailed itinerary for the first year of sailing, so that everyone would know precisely where we would be on a given date and could join us at any time. The method of communication would be simple, by phone or e-mail. Given the cost of satellite phone calls at the time, these were the only possible options. We had only to tell them when we were to arrive at the chosen spots.

My life was thus going to change radically over the next two years. The city of Miami would no longer be for me what it had been in the past. A great turning point in my life had arrived.

When would I see this city again? The broad avenues with many flowerbeds bordered by palm trees covered with lights that called to mind an endless Christmas? The immense new parking lots amid the hundred-year-old trees? The architecturally diverse buildings surrounded by golf courses, whose perfect greens always seemed to have been clipped with scissors? And the poorer neighborhoods inhabited by Haitians, frightened immigrants in search of a semblance of happiness? The handsome houses of the more fortunate Cubans who by work and determination had made a place in the sun for themselves? And the huge freeways crisscrossing the city as they cut through terrain covered with anonymous warehouses decorated with garish multicolored advertisements? I didn't know then that I was saying farewell and not adieu.

I had only to finish packing up my apartment, which I decided to rent out during my absence. It was already the end of June 2000. Time had flown. I was traveling to Europe through New York.

The travel agency I used to book my airline tickets made an offer I couldn't refuse. For the price of a plane ticket, I could take the *Queen Elizabeth 2*, which would sail from New York to England in five days. I immediately saw two reasons to seize the opportunity: first, I would avoid jet lag, and second, I would familiarize myself with the ocean I was to cross a few months later in the opposite direction.

On July 3, my departure date, a friend drove me to the Fort Lauderdale airport with my two suitcases and a piece of hand luggage. The heavy rain typical of the season made my leaving less difficult. The flight to New York had always seemed extremely long to me. Was it because of the lack of legroom or because of the endless procession of states following one another beneath me? The two Carolinas, Virginia, Maryland, Delaware seemed to go on forever before we got to La Guardia. The view of Manhattan was, as always, spectacular. I loved this city from a distance. The forest of skyscrapers, peaceful from afar, blocked the horizon once you were on the ground.

An old friend, Pascale, was waiting for me in her apartment, full of life and happy to see me. Since her apartment on 71st Street between Park and Lexington was not air-conditioned, I opted for a small hotel on Broadway. I told her about the plan for my expedition, and she talked about her work as a psychiatrist. The next day, I filled out a questionnaire from the Bacardi company's public relations firm, to help put together a complete press package that would be distributed to all the media.

I spent an afternoon at the Museum of Natural History, essentially to see a film produced by NASA on the history of the universe, from the big bang to the present. In the course of an imaginary interplanetary journey on board a spaceship launched from Cape Canaveral, we left our solar system and crossed through the Virgo supercluster with a thousand nebulae. When we reached a distant observation post from which the good Earth appeared to be a star lost in the middle of the universe, I didn't know whether we would be able to find our way back.

The *Queen Elizabeth 2* was docked at the western end of 42nd Street, among the docks once used by many steamships. Leaving at night, with Manhattan to port, was a magical moment, a beautiful dreamscape made all the more intense because a group of the most beautiful four-masters in the world, garlands of lights surrounding their sails, were rendezvousing in the port of New York. You could see not the roofs of the countless skyscrapers but a series of strips of little squares of yellow light glittering endlessly in the sky like specks of gold on an evening gown.

Leaving behind the harbor and the mass of golden candles, we finally sailed beneath the illuminated Verrazano Bridge, 1,298 meters long and 220 high, a signal of our approach to the open sea.

Giovanni da Verrazano was the first European to discover New York harbor, in 1524, and he gave it the name

Angoulême. This Italian mariner also had sponsors, including François I and several bankers and merchants of Lyon. Following Magellan's voyage around the world, he landed in China — a sensation at the time — and was given the task of finding a faster route between Europe and the Far East for the silk and spice trade. While looking for this northwest passage on his ship *La Dauphine*, he explored all the bays on the American east coast, including New York harbor, to no success.

During the entire Atlantic crossing I could follow our trajectory on my new portable GPS device. Captain Paul Wright did me the honor of showing me around the bridge, which was equipped with a dizzying array of navigational instruments. When I crossed the ocean, I would not have a fraction as many. Lectures, the exercise room, the sauna, and the pool made the crossing both interesting and very busy. We had calm seas until we were off Ireland, when a force six gale suddenly drove the bow of the ship into huge walls of water. As soon as we were in sight of the southern tip of England, however, the sea calmed. When we reached Southampton, there remained not a ripple.

I took a bus to Waterloo Station, where I took the Eurostar for Paris. I was fresh as a daisy. My two days in Paris gave me time to see my family and to take in an exhibition on the evolution of species at the Museum of Natural History there, another collection of information to put beside what I had gathered from the eponymous museum in New York. I also made sure that money had been successfully transferred, because I would have to pay the last third of the construction cost of my boat, which would be called *Le Prince de Vendée*, simply because I had been born in that region and my family lived there.

I stayed briefly in Brittany, in the Bay of Morgat. With my family, I was able to watch the large sailboat race from Brest to Douarnenez. A few days later, I was on the train to La Rochelle, where my brother Yves was to meet me before I went to check on my boat's progress. In the shed of the Aigrefeuille workshop, the two large white hulls seemed to be waiting for my arrival. The shipyard told me that the boat was a good two months before launch. I had hoped it would be completed by midsummer: now I would have to cross the Bay of Biscay at a very bad time of year. Still, the extra months would allow me to focus on a considerable number of details indispensable for its proper completion.

The last electronic equipment was then selected. Many details of the internal space were decided on. I could modify the superstructure so that it suited me better, suggest new kinds of storage, elevate the berths, have the sheets fitted to their size, and complete the diving equipment — a thousand tiny details that, I hoped, would make a difference in the comfort of the expedition. I designed our flag and had it made. And I did not forget to take all my drawing and painting supplies. I took a quick trip to Bordeaux to see a friend, Béatrice, who was to join us on Barbuda to be our cook.

It was not until mid-September, ten days before launching, that I could load on board all the boxes that I had had sent from the United States. The work was facilitated by the dozens of kilometers I covered by bicycle between my brother's house and the port of Les Minimes. I decided to take on as crew members Patrick, former director of the port of Papeete, and a friend of his, a doctor in Fouras; another friend from Listrac would meet us in the Canaries. It was now mid-October and we were waiting for a favorable weather window to set sail. In the meantime, we filled the diesel tanks and conducted the first sea trials to familiarize ourselves with this new boat, which gave us the opportunity to rectify the mounting of a sounder that had been improperly done. On October 18, 2000, we sailed from La Rochelle at four in the afternoon, in a harshly cold atmosphere and under a driving rain. At around seven, we passed the Chassiron light at the northern end of the Île d'Oléron. An east wind had already produced high seas. We were heading for the island sun. The boat on which I found myself — my boat — made of fiberglass, steel, and nylon, bore a strange resemblance to a little ship I had built out of wood with paper sails when I was a child. I sailed it in a stream that ran between two hills that divided my village. Each of us — the boy and the man — has a boat in his heart.

3

The Clock of the Atlantic Ocean

La Rochelle in October — The hell of Biscayne Bay — Lisbon and Lagos — Henry the Navigator — The dead man of the Canaries — The first volcanoes — Preparations for crossing the South Atlantic — Books on board — Listening underwater — Plunging into the ocean — Our encounters at sea — The other side of the world.

At two in the morning on November 23, 2000, we were 125 miles south of Grand Canary. "Rouge sur rouge, rien ne bouge."* Complete calm. A freighter heading north very far off the African coast showed in the distance its red port lights.

The huge gray-black mirror of the surface of the ocean clearly reflected a troop of thick, dark, and worrying clouds. Ahead, the line of the horizon was hidden like a ghost draped in black wanting to draw us toward a cliff of uncertainty. A

*"Red on red, nothing moves." A maritime expression: when the sidelight on your boat is the same color as that of the other boat, there is no risk of collision.

few patches of sky here and there gave us a glimpse of the immense vault of glittering stars. The *Prince de Vendée* was wending its silvery way through this seascape without a single breath of wind. Only the purring of the engines graced this moment outside time. I was finishing my watch at the helm, ready to turn over our fate to the next crew member. The aroma of fresh coffee wafting deliciously as far as the rear deck largely covered the faint smell of the sea breeze created by our movement through the night. We were still far from the tropics. The freighter, like a silent cat sliding over the glassy surface, had noiselessly disappeared a few minutes earlier, leaving barely a trace of its brownish smoke on the horizon and this steely surface.

Everything had not been so majestic and calm since we had sailed from La Rochelle. The day after starting out, we officially entered the Bay of Biscay. Its notorious reputation, of which we were all aware, particularly at that time of year, preyed on our minds. By the second day of sailing, after the calm morning following the strong southeastern swell that awaited us when we came out of the Charente channel, the wind shifted quickly to the northwest. The barometer had fallen during the night. A heavy swell indicated the proximity of the depression that had grown stronger over the North Atlantic. The two reefs taken in the mainsail and the jib were no longer enough to stabilize the boat; the waves were mounting moment by moment, and we were going to have a bit of a struggle. Our boat was being battered. The sea roared beneath our two hulls with a deafening noise. I had difficulty standing at the helm, and, looking at our wake, I had the impression that we would never get across those mountains of water. After a delicate maneuver, we were now without sail and depended entirely on our two engines.

A Spanish launch twenty-five meters long, not far from us, whose red hull was forcefully cutting through the ocean's fury, disappeared from our view for interminable minutes. We would have preferred a calmer baptism. Our new hulls were severely tested and their adjustment to these uncontrolled movements created streams of water swept by large waves around the bridge portholes. It wasn't long before there was saltwater in the cabins. Some of us drained the water as quickly as possible while others very calmly tried to hold the boat against the waves, like rodeo riders grasping the flowing manes of wild horses. We were in a force eight gale, with winds at forty knots. The cresting tops of waves six or seven meters high crashed against our hulls, giving off great bursts of spray that spattered with force against our faces. It was hell. Thank you, science . . .

We were in sight of Cape Finisterre just off the bay of La Coruña. Barely were we past the cape when it seemed as though those moments of horror had never existed, and the Portuguese trade wind took us under its gentle wing up to the mouth of the Tagus. A calm night quickly quieted our fears. Hundreds of boats sailing for Europe under the starry sky passed us a few miles from the coast. Their lights, which we attempted to decipher in the distance, were another reason for us to stay alert. Our journey was finally beginning. We had hardly had the time to notice until then. Everyone was finally happy on board.

Entry into the port of Lisbon early in the morning was rather tricky because of the many buoys indicating the channel to follow, but once past them, passage posed no major difficulty. The old tower of Belém, recalling the exciting maritime history of the city, was lit by the golden rays of the rising sun. The huge red bridge indicated the location of our marina.

A visit to the Maritime Museum could not be passed up. It was there that I learned that early in the fifteenth century Henry the Navigator called on the Moors then occupying the southern part of the country to fit out his fleet. Only they knew the secret of the trade winds. This was no small matter at the time, because the information was considered a state secret. We took advantage of the layover to repair the slight damage on board and to visit a local branch of Bacardi, which presented us with a large number of bottles of rum.

A few days later in the late afternoon, we resumed our journey in the direction of Lagos. After an uneventful night, we found ourselves very early the next morning off the Cape de São Vicente. This promontory topped with a red lighthouse overlooking the famed maritime school established by Henry the Navigator had already seen off such great sailors as Vasco da Gama, Christopher Columbus, Magellan, and Bartholomeu Dias, who was the first to sail around the Cape of Good Hope in 1487, followed by many others. It was in this school that the first plans for the caravel were conceived, along with the first navigational instruments, many maps, and new methods of navigation.

It was October 27, and Lagos was not far off. We had to take on supplies before heading for the Canaries. Three days later, aided by a constant northeast wind, we reached the gleaming marina of Puerto Calero on the east coast of the island of Lanzarote.

We could not miss a tour of the volcanoes. This was my first encounter with the unusual world of this planetary phenomenon. Visiting the Timanfaya National Park located in the middle of the island, I became aware of the power of the heat streams rising directly from the center of the earth. The Canaries volcanoes belong to a category known as "hot spots," the eruptions of which do not in any way derive from rising magma caused by the friction of tectonic plates at great depths. In these islands, the upsurges of magma come from the center of the earth, which releases bubbles of matter. These bubbles rise slowly and explode on the surface, with the depth of the departure point depending essentially on the geological composition of the magma. The last major eruption took place in 1730, lasted for six years, resulted in the creation of thirty craters, and covered the entire southwestern part of the island in lava. This flow, sometimes as high as ten or fifteen meters, is still visible today. A brief attempt at scaling a slope quickly reminded me of the dangers of this fused rock, which in cooling had changed into stone as sharp as steel.

I retained an indelible memory of this island. Seen from a distance, it looked like a huge black ribbon over which had been scattered charming little white houses with red roofs. The island has made enormous efforts to protect its environment, with an artist named César Manrique at their source. Despite the season, it was already very warm, and the next day we all decided to go swimming on the leeward side of

Lobos Island, adjacent to the north coast of Fuerteventura. This was our first swim in the ocean. Some fishermen invited us to share a meal in their shelter on Puertito. The water was very clear and tempting for me.

A few days later, I decided to test all my diving apparatus in a little inlet of Fuerteventura known as the Tortoise, because of the rounded form of the deserted hill overlooking it. The bottom covered with gray sand had no vegetation and no life. Swimming underwater toward the northern part of the inlet and just at the base of a small cliff, I found a startling number of old T-shirts scattered around. The idea came to me of taking a few to give to my crew. Sighting one that was brightly colored and might make an attractive gift, I had some difficulty extracting it from the sand in which it was half buried. To my surprise and horror, I pulled out with it a piece of bone that I immediately identified as a human tibia. To confirm this macabre discovery, I pulled some more, and the femur appeared. Frightened and distressed by this discovery, I hastily returned to the boat to recount my adventure.

The problem then was to decide whether we would inform the competent authorities, with all the difficulties that might create for us. I then recalled a television program that I had seen in the United States describing large-scale human trafficking between the African coast and the Canary Islands. On the coast of Mauritania, at places only fifty-four miles from this spot, unscrupulous individuals built barely navigable makeshift boats, enabling them to extract large sums of money from Senegalese men for a passage to Europe through these islands. The skeleton must have come from one of the many wrecks that this traffic produces. We decided, perhaps wrongly, that that spot would be his final resting place. We said a silent prayer for him, imagining the suffering of his final moments.

Since we were ahead of the scheduled date we were to pick up Michel, our new crew member, at Grand Canary airport, we decided to stop at several small ports and shelters, such as Gran Tarajal, Morrojable, and Jandia Point. I was surprised to find that the menus displayed outside all the restaurants on land were in German. There were also Bavarian inns serving German beer.

We were now 120 miles from Puerto de Mogán, the marina at the southern end of Grand Canary, which would be our last stop before the open ocean.

After sailing by the majestic southern cliffs and being in sight of the marina's dock, we discovered that we were not the only ones to want to make the crossing. There was no room for us at the dock; most of the boats were taking on provisions for the great crossing, and the pontoons were swarming with dozens of sailors searching for the last supplies they needed.

We were forced to moor outside the little port, at a half cable from the nearest pontoon. Our plan was the same as everyone else's. With the help of an indispensable list that a friend had given me in La Rochelle, enumerating everything a person would need for at least a month at sea, we had only to multiply by the number of people on board to figure out what had to be bought. Three days later we were ready. We had filled the reserve tanks with 350 extra liters of fuel, which made us ride a little lower in the water. What we had not taken into account was that some crew members had also taken on a little too much in the way of liquid supplies: when I discovered that we had on board more than 150 liters of wine, it was already too late. That was four bottles per person per day, considering that I drank no alcohol. As we shall see, that overload turned out to have some bearing on our safety.

At 11:45 on November 21, with all sails unfurled, we left our mooring. We had a northeast wind of sixteen knots, and already, a few miles offshore, our *Prince* was sailing at a good clip. A little anxious, but very excited, we gradually saw the steep coast of our last island fade in the distance. A short distance to starboard, a school of dolphins wished us a fair wind.

We had covered 1,610 miles since leaving La Rochelle, with 2,710 more to go till we reached Antigua. The sky that morning was as blue . . .

Every three hours during the crossing, we noted in the ship's log the wind speed, the state of the ocean, the visibility, our geographic position, and any comments on these observations.

The division of tasks was well established from the beginning, an arrangement discussed by all, and was posted in the wardroom. The principal elements were the scheduling of night watches, cooking, dishwashing, housekeeping, and keeping the log. Only maneuvers that had to be explained in detail by the captain before they could be carried out were done in common. Setting the course was my responsibility. I had learned from experience that only a flexible but quasi-military discipline functioned well on board because of the enforced lack of privacy. Often a casual attitude established at the beginning out of friendship resulted in serious internecine disputes in a crew. The result was not perfect harmony among us, but at least we were able to deal with one another like civilized people. A few months later, in a conversation with some other amateur sailors, I was told the story of one crew member who ended the crossing locked in his cabin at gunpoint.

The drought that had affected the Sahara since the month of August led me to believe that high-pressure zones would be located very far south, close to the equator. In order to avoid the calm that they would produce, I chose a route farther to the north than that recommended by the charts, thus not heading as far south as the Cape Verde Islands. Our first destination was the junction of the 17th parallel north and the 32nd meridian west. Then, depending on our position , we would take the best route to reach Antigua, located at the 61st meridian west. On the basis of these decisions, I hoped to meet up with the trade winds as quickly as possible.

For those enamored of the sea, we steered that night at 270 degrees, we were at the 22nd parallel, the night sky was brightly lit, and we seemed to be moving at the same speed as the millions of stars above us. Familiar with the night sky, I easily located the Southern Cross, the Big Dipper, Leo, Cancer, and Virgo, constellations we don't take the time to admire when we are on land. The feeling of extreme solitude that one might experience in such circumstances, in the middle of the open ocean, is dissipated as soon as one looks at the sky. Even if there is no life outside our galaxy, the 200 billion stars that populate it glittered sparklingly that night to remind me in a strange way that I was not alone in the world. The dark roof pierced millions of times with constant lights gave me the impression that I was in a tent, sheltered from harm, and that from above, a thousand attentive eyes in that diamond-studded vault were watching over me.

For several nights during my watch, which ran from two to four in the morning, I tried to locate a star named Altair, which is in the middle of the constellation Aquila in the southern sky. Unfortunately, we were too far north of the equator for me to see it. *Altair* was the name Henri de Monfried had given to his second boat, built in Djibouti on the beach of Obock in 1920.

I had brought a number of books with me. I wanted to use my free time during the crossing to read about the lives and achievements of Bougainville, La Pérouse, Cook, and the master of the secrets of the Red Sea. I was surprised by the details of Bougainville's description of the work of the Jesuits in Paraguay, and of course by the colorful way he described the customs of Tahiti, as well as his generosity in outfitting a ship at his own expense so that poor Acadian families driven from Canada by the English and then living in Saint-Malo could find refuge on the Falkland Islands. As for La Pérouse, a courageous native of Albi, I admired his decision to fight alongside the Americans in their war of liberation, and I envied his astonishing voyage to the Pacific accompanied by the most learned hydrographers and naturalists of the time. Despite the sad end of his frigates, *La Boussole* and *L'Astrolabe*, on the reefs of Vanikoro in 1788, he left behind invaluable work on currents, sea temperatures, and the hydrography of the Pacific Ocean. Finally, there was the nerve of Monfried and his love of hair-raising adventure, his turbulent life on the coast of Yemen, his gilded prison in Kenya, and his unusual end in his house in Berry. My fondness for this character brought me a few years later to visit his houses in La Franqui and Ingrandes. I had also brought with me a biography of Prince Albert I of Monaco, and, borne by the waves, I passionately followed the scientific and oceanographic work that he had done on board his boats.

Among our activities on board — we had to do something to pass the time — one of our favorites was deep-sea fishing. Using two large rods and reels at the stern, we regularly added to our lunches and dinners tender fillets of mahimahi. The only explanation for their presence hundreds of miles from land was that they were looking for food, so the sea was populated everywhere. Life in the middle of the ocean surprised me more than once — we also encountered birds twelve hundred miles from shore. Then there was an encounter with a gray razorback whale about fifteen meters long and weighing a good two tons, which amused itself for an entire day by slipping between our two hulls, risking a collision. Then it happened, he touched us, the jolt knocked down some dishes, and we thought for a brief moment we were in real trouble. But with a flip of his dorsal fins, he left without making excuses. A quick dive under the hull showed me that he had done no damage. Our *Prince* was solidly built.

We had decided from the outset that once we encountered the trade winds we would set our sails as follows: during the day, we would unfurl one of our two spinnakers, with a surface of 140 square meters, one of which was provided by our foundation, and at night we would use the 47-meter jib for greater safety in case of a sudden increase in wind velocity. We thought that this was the best way to cover the miles. And that's what we did every day — except one. That afternoon, after playing our daily hand of cards while listening to our favorite music, it was time to change sails for the night. The setting sun was once again opening its orange curtains to give us a free show. The atmosphere on board was superb. Lunch had been washed down with plenty of wine, too much wine, followed by cognac and evening aperitifs — all of which had put my crew in a trance of laziness and bliss. When it was time to get to work, any argument was used to avoid it: the weather was fine, there wasn't a cloud on the horizon, the barometer was steady, we would travel faster. All my good reasons for doing what we were supposed to were swept away in the general euphoria. I allowed myself to be persuaded for once, thinking that I was wrong, but also that it was right to give in on occasion. We could all drink to that again.

Some had no trouble falling asleep that night. But we hadn't taken account of the tricks of the weather. Around midnight, wind velocity started to increase, quickly reaching thirty-five knots, and since our spinnaker, made with sixty-five-gram nylon, could withstand only twenty-five, we had to move quickly to take it down. While it was very easy to take it out of its case by oneself in good weather during the day, that night I was unable to take it down without help, as the boat was moving in every direction at once and the sea was growing heavier by the minute. The maneuver, done with a crew that was barely awake and still under the influence of alcohol, was a disaster. Not only was it hard to bring the spinnaker down onto the deck because of a total lack of coordination, but we lost one of its braces, a fifty-meter line that disappeared in the night because of a misjudgment of the speed of the maneuver that had to be made.

The jib was quickly unfurled to stabilize the boat, which went on its way in the dark. The next day at dawn, the sea was still heavy, with waves of two to three meters. When I got up I saw our line floating to the stern. It took me two hours of work, with a rope attached to my waist, underwater, with five thousand meters of depth below me, battered by the waves that continually brought the hull down on my head, to free the spinnaker brace, which had been caught in the port propeller in its nocturnal flight. It provided a good lesson to encourage my crew to drink in moderation and to keep the rest of the wine cellar for their landing.

Who stole the trade winds? For three days we had been drifting with only a little lapping in front of the boat telling us that we were actually moving forward. We still had a long way to go; our fuel reserves would not last forever.

It was not until we reached the 24th parallel that the winds finally showed up, with their twenty-knot force driving us at a speed of seven knots; the barometer read 1015 millibars, and we had just entered a different time zone. It was November 24; it was very warm, and Europe was far behind us. Every day we listened to the weather report of Radio France Internationale and carefully noted down the depressions in the Alice, Josephine, Meteor, and Madeira zones, in order to determine their effects on the Cape Verde zone in which we were now located. All our careful calculations and forecasts had the advantage of keeping us busy, but that was all. The day our trade winds showed up was the day when one of us said, "I told you so," which shows you how much we knew.

How those trade winds made us dream. They are hidden in many other places with other names. The south winds of Long Island and Ireland, the north winds of Portugal and the Canaries, the east winds of the West Indies. They turn and turn, creating currents that turn as well. They carry saltwater, freshwater, cold water, warm water, clean water, dirty water; the water of the Orinoco mixes with the water of the Hudson and then picks up the water of the Loire, and it all turns and turns around the Atlantic Ocean like the hands of a giant clock.

Imagine for a moment that you could color all that water and then set huge red buoys at intervals of a few hundred miles, and set two giant arrows, one larger than the other, in the middle. We could indicate the time to millions of other planets. Even if the time was wrong, it wouldn't matter; we would have the satisfaction of being superior clockmakers. That's the kind of aberration that your mind drifts into when you have no wind and you have to dream about its appearance. Sometimes time hangs heavy on board and minds wander.

During the crossing, we saw no one, apart from the freighter along the African coast. But we spoke frequently with sailboats that we could not see; one of them was kind enough to call our families to tell them where we were, another, in a great hurry, told us that he absolutely had to reach Antigua by the following Saturday so he wouldn't miss the holiday celebration held that day and still another asked about the wind where we were, because he was complaining of the doldrums in his area. No light at night, nothing, the world was unpeopled. We were all alone.

At two hundred miles from our destination, our boat was completely covered in salt. We hoped, despite the large cumulus clouds constantly hovering above us, that we would have a little rain to wash off our hull, but none came. The drought that had appeared over Africa was obviously pursuing us.

A cross-swell of unknown origin shook us mightily for two days when we were a hundred miles offshore. A few days before reaching Antigua, there was heavy betting about our date of arrival. Would it be the eighth, the ninth, or the tenth of December?

It was in fact December 11, 2000, at dawn, that the island of Antigua gradually emerged on the horizon directly before us. First there was a slight shadow, then a strip of land barely burnished by the rising sun, finally beautiful hills colored Caribbean green. The crew were already making plans to fly back to their families in Europe.

It was just about noon when we anchored in the little bay facing the marvelous port, fortified by Admiral Nelson, known as English Harbour.

In all, we had covered 2,738 miles from the Canary Islands, and our log indicated 4,340 from La Rochelle. Our average daily distance had been 155 miles during the eighteen days of crossing, at a speed of 6.5 knots.

Now everything could begin.

4

The Thousand and One Surprises of Barbuda

The welcome at English Harbour — First contact with the government — Costly moves — Wind at the dock — Change of base — Recruiting a new crew — First steps in sight of corals — A tricky arrival — The first wild mooring — Discovery of the island — Joseph and John — The local government — The fishermen with us — The loss of a boat — A visit to church — Work in the mangroves — Signs on the cliffs — Listening to fish.

As he did every morning when he was going fishing, Joseph Vernant got up early. After drinking his coffee and eating his eggs, he went out of his little house with its ill-fitting tin roof and its wooden walls covered by faded and cracked pale green paint. Opening the wire gate that squeaked on its rusted hinges, he was so used to the morning noises that he didn't hear the piercing, staccato crowing of the roosters in the darkness or the mocking braying of the donkeys in the distance. He did notice a bit of white smoke from a smoldering brushfire floating above his garden, which was strewn with old cassia plants and torn netting. He thought only about what the weather would be like at sea. He lived in the middle of Codrington, the only village on the island of Barbuda. He got into the old pickup truck parked on the road outside his house, onto which he had just loaded his diving equipment.

Once past the long, straight stretch lined by houses identical to his, you came to a barely paved road that gave way to a sandy track full of ruts. Only the electricity poles bordering a scanty pine forest reminded him that, even a few years ago, the island had still had no electric power.

Dawn was already breaking when he came in sight of the sea. His boat was waiting for him on the beach not far from the old Martello tower that had once been a fort, built by the first inhabitants of Barbuda. The sea was fine, it would be a good day.

Joseph was robust, and his bright eyes constantly mirrored the blue of the sky shadowed by the whiteness of the clouds that must have drifted before his sight for years.

In order to avoid friction between fishermen, the local government had divided the island into two territories — the south where Joseph lived, and the other north of the capital including the lagoon and the coral reefs off the coast. These two large stretches of ocean could feed the whole population.

Like all the other fishermen in the south, Joseph knew every corner of the area, from Palmetto Point to the "Spanish" Point, all the lobster holes, all the shallows and the coral reefs that surfaced at low tide, all the bays surrounded by rock, all the overhangs under which grouper and porgies lay concealed. He had overcome all the dangers hidden behind the Palaster reef and faced the fierce waves that swept the coast as far as Pelican Bay. He had already lost two of his nets and several lobster pots in the storms that always came from the east in that spot. He knew how to find the good places to fish, depending on the weather. Indeed, his livelihood depended on his knowledge of the area, which had enabled him to earn an honest living, replace his outboard motor every four years, and repaint his boat, which had a pointed bow and a square stern. Once his boat was in the water, he pulled the cord to start the motor and sailed off with a smile on his face, relishing his freedom.

John Iudel was a fisherman from the north of the island. He still lived with his father in a modest house at the side of the road leading to the "Highlands," the northern cliffs. His boat, more modest than Joseph's, was sheltered in a little muddy inlet in the lagoon, very close to the village. He was younger, although that was hard to tell, so he had less experience, but he took more risks. He could have followed the way out of the lagoon at night with his eyes closed. Avoiding the mangrove roots, following the good channels that changed direction with the tides, sailing very close to the coral barrier that ran from north of Goat Point to Boar Point, finding the least dangerous channels, and then sailing into open water and finding good spots to drop his lines — all came to him easily. Only once had he been caught by darkness and bad weather, when his motor had stalled and he had not been able to start it again. All his fishermen friends, worried at not seeing him return at a reasonable time, had gone looking for him.

These two strong and courageous men, of contrasting character but both extraordinarily generous, were to be our guides and became our friends. Both always wore old shirts full of holes. When we regretfully left them six weeks later, they proudly put on the new polo shirts designed for our expedition.

Many things had happened since we had reached Antigua. After the first day, I decided to change our mooring, in order to be in a quieter marina where I could seriously tackle the work that had to be done after those two months of transatlantic sailing. For the next few weeks, Falmouth Harbor would be the perfect place to finish the work and also take advantage of the sun and the luxuriant landscape that spread out to the rear of the boat. I wanted to finish everything before Christmas. A few employees of the marina helped me through three days of thorough cleaning, from the hold to the bridge. My *Prince* smelled like a new boat. Royal poinciana blossoms decorated the wardroom table.

New oil filters were installed in the engines; the bottom of the jib, torn by constant rubbing against the base, was restitched; water supplies were replenished; a new anchor with a longer chain installed; the hull thoroughly scraped; the bases of the shrouds tightened; and a thousand other tasks accomplished, all of which took a full week. I am ashamed to say it, but I was very happy to be pretty much alone on board.

I had to face one problem in the middle of the night when fifty-knot winds howled through the marina and my two anchors started dragging. The power of both engines stabilized the boat at a good distance from the dock until morning.

I informed all my contacts by phone and e-mail of my safe arrival in Antigua and told them of my planned departure in mid-January for the first research work in Barbuda.

In the meanwhile, local representatives of Bacardi had organized a press conference in the best hotel in the southern part of the island, the St. James's Club. The president of Antigua, Lester Bird, Environment Minister Molwyn Joseph, and Shirlene Nibbs, minister for tourism, were to attend with a number of other local notables. The local press and television would also, of course, be there. I therefore had to prepare a presentation of the research program of our expedition in their territorial waters. Before leaving Miami, I had prepared a series of slides presenting our work, which made my task a good deal easier.

That evening a few hundred people crowded around the sumptuous buffet in a room decorated with our spinnaker bearing the colors of the foundation. Facing the Bay of Mamora, the hotel was a perfect setting for this kind of reception, made even better by the setting sun illuminating the emerald waters of the little cove next to the beach.

Introductions, as in all the islands formerly occupied by the English, followed strict protocol. I met everyone, made my presentation, and answered a number of questions. To judge by articles in the press and television reports, it was a success. In the course of the evening, I made appointments with various officials in order to obtain the necessary authorizations to go to Barbuda in January. I also met Pierre, head of the photo service of Agence France-Presse in Paris, who was on vacation with his wife. He later became both a friend and a strong supporter of our expedition.

The next day, I took a bus to the island capital, St. John's. Coming into the modest suburbs and very near a collection of little huts that served as a vegetable market, I was surprised to see a shiny new building. With the latest in freezer compartments, water faucets facilitating thorough cleaning, and neatly aligned cement stalls, this was the fish market. A plaque at the base of the building indicated that this was a gift of the government of Japan to the island of Antigua. Since we intended by our soundings to try to locate the mating zones of some species of fish, I hoped that this would not complicate our work.

I knew that votes in the International Whaling Commission were very costly, but I had no idea how costly. Well, now I knew. I also hoped that this was the only concession made by the government.

A few days earlier, a neighbor in the marina, a big-game fisherman, had told me that he had not caught any tuna for several years. I therefore wondered whether Japanese trawlers had been given authorization to spread their huge rotating nets in these waters.

Entering a crowded government office, I was greeted by a friendly secretary who had been informed of the purpose of my visit. Half hidden behind his desk covered with mountains of files, he informed me that the fishery service located not far from the new market would give all the papers necessary for my work in Barbuda.

When I got there, the place was deserted, there was not a single paper on the green metal desks, and a solitary fan was humming on the ceiling because of the heat in the room. Finally, after I had waited an hour, a man wiping sweat from his brow introduced himself to me. In the absence of his superior, Cheryl, who was on maternity leave, he was the only person able to grant me the authorizations, provided he received a fax from Cape Canaveral confirming our mission. The letter of accreditation from Professor Gilmore addressed to his government, which I showed him, was not sufficient.

An entire week, a week far too long for me, was spent in discussions with this bizarre individual. I had to take the bus, and on each visit I lost valuable time. During that time he received the requested fax, but that still wasn't enough. He now needed more papers. I soon understood that the kind of paper he needed had nothing to do with our work and everything to do with his personal interest.

Fortunately, at my last visit, his boss had returned. A charming woman, who had just become a mother, she greeted me warmly, even begging pardon for her absence. Extremely interested in our upcoming work, in my presence she called a representative of the local government of Barbuda; everything was organized for our stay in a few minutes, and she gave me a letter authorizing a stay of two months on the island, which she adored. As I left her office, I thought that the administration here really needed people like her. Thank you, madame, and Merry Christmas to the baby.

The next day we celebrated Christmas with some neighbors in the marina, feasting on the traditional turkey, prepared by some people from New Zealand passing through Antigua.

That evening we went to a Christmas service in an Anglican church. This little chapel was perched on the hillside just above the marina. All the inhabitants of the village — men, women, and especially the children — were dressed in their finest.

When the hymns, to the accompaniment of guitar, saxophone, and drums, rose in the air, with candlelight illuminating the simple polished wood vault, we were all deeply moved. The long threnody of warm and rhythmic voices and the languorous swaying of bodies accompanied by clapping in time with the barely syncopated rhythm of the little orchestra was all too much for me. Hesitant tears ran down my cheeks, and I was not alone. We were all thinking of our absent families. I wished myself good sailing.

The next day, after saying my good-byes, I left the marina, alone on the boat. I had an appointment for a few days later with my new crew in a port in the northwestern part of the island whose location would shorten the distance to Barbuda.

As I left Goat's Head Reef to port and rounded Johnson Point in the clear morning light on a calm, transparent sea, I sensed that the real departure was very near.

Jolly Harbour was exactly as it had been described. The entry, well protected by two large hills, opened onto a veritable interior lake bordered by charming houses. The wooden docks, arranged to provide some degree of privacy for everyone, were well equipped. A small shopping center in the handsome architectural style of the islands was near the docks.

To crown the idyllic character of the spot, the next day I discovered that next to this little paradise was one of the most beautiful white sand beaches I had ever seen. I had about ten days before sailing, and the wait would not be painful.

However, having learned that there were no supplies available on Barbuda, I now had to take on board enough food for two months.

Simon, a young American I had found through an employment agency, arrived punctually from his native Massachusetts. He turned out to be by far the best crew member that *Le Prince de Vendée* ever had. Not only was he a good sailor but, because he had worked in a reputable shipyard at home, he was very knowledgeable about mechanics and sails. His competence, discretion, and willingness were appreciated by everyone during the three months he spent on board.

The final preparations were interrupted by pleasant moments of relaxation. New friends made the time seem too short. Dominique, a woman from France, prepared sumptuous meals for us in her house near the marina; Roger, a Canadian who owned a superb sloop, was interested in our research. He wanted to join us later and did meet us at our next port of call, Saint Barts. He often invited us for an evening drink on his boat. Of course, we too had parties on board. On New Year's Eve, before the fireworks display put on by the local hotel, we all watched from the beach a sunset that I will long remember: against a background of red and golden streaks lay the shadowed outlines of the smoking island of Montserrat, the stony island of Redonda, and the majestic volcano of the island of Nevis, all reflected in the calm bluish orange surface of the sea.

I took advantage of this period of calm to keep in shape, swimming every morning, diving to test the equipment I would need later on, and going out in the dinghy to make sure the hydrophones worked properly at sea.

We were completely ready on the day Béatrice arrived from Bordeaux. As we had arranged, she would have the burdensome responsibility of cooking on board for a month. We had no idea that she was such an accomplished chef, and after she left we sorely missed her marvelous apple pies and lobster gratins.

The next morning at six, January 9, 2000, we sailed for Barbuda. It was absolutely essential that we arrive before noon. We thought that we would not return to Antigua before the end of the year, but the events that took place during our stay on the little island to the north were to prove us wrong.

Barbuda is entirely surrounded by coral reefs. Navigation in the area is extremely tricky, and for that reason few boats venture into those waters. Our twenty-five-mile crossing was faster than we had anticipated, thanks to a strong wind on the beam that pushed the *Prince* to its limit, sixteen knots. With the sun at its zenith, we had excellent visibility to navigate between the coral reefs.

We had to go around two imposing reefs to our starboard, and then line up with a tower and the top of a hotel roof at Cocoa Point to find the right channel leading directly to our anchorage. From my perch on the mast, it was easy for me to point out to Simon the best place to drop anchor. The depth was four meters, and we were half a cable from a beach of pale pink sand, halfway between Cocoa Point and Spanish Well Point. We were just to the south of Barbuda and would be there for some time.

Beyond the beach, which stretched for several kilometers, we could see the tops of the mangrove trees. Tomorrow was another day, but for now the only thing was to get into the water.

We soon contacted the local authorities through our radio on board. A van would come to pick us up early the next morning, and Simon would come with me.

The village of Codrington, the charming capital, was located at the junction of the main road and a track leading to the huge lagoon bordering the village on the west. A mixture of little multicolored huts and more substantial cinderblock houses surrounded by vegetable gardens gave the spot a good deal of charm. Children on bicycles and women at their thresholds in hair curlers waved at our driver. A few shops gave glimpses through their modest windows of stacks of brooms and canned goods.

It had taken us an hour to get there, traveling mostly through a landscape of brush, mancenilliers, spiny shrubs, and pine trees. The badly maintained road was deserted that morning. The driver told us, however, that he had to be careful because wild pigs and deer frequently crossed the road. Acting as a good guide, he also told us that once a week trucks loaded with sand drove to the little pier to discharge their load onto a large barge that carried it to islands to the south. There really was a sandman. But this trade was not all to the good, because it had negative effects on the environment of the island, he told us.

We had come to an area in which several government buildings were scattered around. A cement staircase took us to the second floor of one of them. There, in a large room with whitewashed walls, six men seated behind school desks set in a semicircle were waiting for us. It was an important moment; all of them later became our friends. The members of the council of fishermen, chaired by Peter Frank, were all present. Following a warm welcome, handshakes all round, and a brief but very formal presentation, they all asked us to join them in prayer. Heads were bowed over the desks and one of them began to sing a hymn. I was prepared to talk to them about microphones, hydrophones, protection, preservation of species, NASA, evolution, science, and the like, and they were communicating with God. Maybe it was all to the good.

After a few minutes, in a very natural manner, they asked me what I needed to accomplish my work, about which they had already heard. But I explained to them, probably giving the false impression that I knew more than I did, the importance of knowing the sounds emitted by certain species in order to be able to protect them, the difficulties that we might encounter, and so on. These were special moments, and we were all speaking the same language, man to man. They accepted our invitation to lunch on board the next day, thanked us for our presentation, and offered a final prayer for our success, in which we all joined. Leaving this moving meeting, perhaps touched by grace, I knew that we would do good work on Barbuda. Joseph would be our guide for the south and John for the north.

The next day, after a superb lunch prepared by Béatrice, they all became just like our brothers. They had all come, Bungy, Elwyn, Moose. Their simplicity amazed me. We agreed to meet regularly, so I could tell them about the work that had been accomplished. In case one of the two fishermen was not available to accompany us on any particular day, they would provide someone else, and it would be up to us to pay him for his time, an arrangement we gladly accepted. All we had to do was set to work, which we soon did.

The sea was running high and we were no longer in the buoy-marked lagoon of Cape Canaveral. We would have to fend for ourselves; we had only a short time to get to know the place — the moon would be full on February 9.

On each of the following mornings, accompanied by Joseph, we went in search of the places where he thought the species that interested us would be found. We either recorded them on the very detailed map that we had, using our portable GPS device, or by dropping little buoys painted bright yellow at the good spots Joseph showed us. We then dove immediately to verify that the fish beneath us were familiar. If they were, we returned at nightfall with our listening equipment. One day, testing our microphones, we had Joseph listen to the sounds emitted by some porgies. What he heard through the earphones made his eyes pop. This first little experiment taught us two things: on the one hand, we were surprised to hear their dry clicking in the morning, because we had been told that they emitted sound only at nightfall, and on the other, the noise of the waves breaking against the hull of the boat interfered with clear recording. We could therefore use his boat only for work in a calm sea.

The huge stretch of ocean was divided into three parts. In one, near our boat, we could work alone in the afternoons with our rubber dinghy. In another, farther off toward the coral reef, we would work at nightfall. The third, more distant and more dangerous, could be covered only with Joseph's boat. But for now, we knew enough to begin. On each excursion, we noted the species recorded, the exact location, the temperature and salinity of the water, the estimated depth at which the microphones were set, the state of the ocean, and the method of research. All this information would be attached to the minidisks we sent to Professor Gilmore in Florida.

During this period we lost many of our buoys to bad weather, replacing them with pieces of cork we found on the beach, which we tied with nylon cords to rocks heavy enough to keep them in place. Our night excursions, near the coral reef — especially when the wind freshened late in the day, were often very risky. I remember one night in particular when we were wildly tossed about and got totally lost in the labyrinth of madrepores. Afraid of puncturing our little rubber dinghy, drenched and exhausted, far from the coast, I wondered if all the work would one day be rewarded if we came back alive. I took us three hours to get back to the boat, in complete darkness under a driving rain. Béatrice, who had stayed alone on board to prepare dinner, had tried to radio for help without success. I need not tell you how much we appreciated her meal. And in fact our recordings that night seemed to be accurate.

During the many dives that enabled us to locate our happily chattering fish, we had all the time we needed to study in detail the different coral reefs to the south and east of Barbuda. We detected no disease in the white strip, no necrosis, no trace of infestation, even less of bleaching caused by elevated temperatures preventing the symbiotic algae located in the tissues of the polyps from fixing the coral stone —

nothing of any of that. The reason for this was simple. Most madrepores, whether branching, in the shape of umbrellas, or in that of cushions, had been heavily damaged by the 1999 hurricanes. We observed a large portion of the staghorn coral, the coral swells, the lettuce coral, the acropora, and the stylophora lying on the bottom, sometimes forming huge piles of limestone. Here and there, however, life was recovering, and a few polyps were growing again. For these organisms going back 500 million years and growing one centimeter a year, the situation was sad but not desperate. All this information was sent to NOAA in Miami.

Unfortunately I knew nothing of ethology, the study of animal behavior. Nonetheless, I was able to observe many things that I found extraordinarily interesting, for example, detecting the fishes' territory and guessing at their various defensive reactions, witnessing their life in schools and alone, noting how some of them found food, wondering at the habits of little schools of young fish, watching them dart into their favorite hiding places at the slightest danger, and finally becoming aware of the interaction of some that seemed to live in couples. After seeing all that, I was no longer surprised that some of them had the skill to emit sounds and perhaps to communicate with one another.

Joseph asked me several times about the source of the sounds. A muscle known as the sonic muscle is located at the top of the air pocket enabling the fish to maintain balance, and a contraction of that muscle is enough to push a wave through a series of small bones to the inner ear where it is then turned into sound. I drew a picture that seemed to satisfy him, and he told me that he would talk about it with his girlfriend, a schoolteacher, so that she could discuss it with her students.

During those two weeks, we had managed to locate and record a dozen categories: some Nassau groupers, grunts, cutlass fish, blue chromis, yellow-tailed sardes, the sound of fire parrotfish nibbling the coral, sabre squirrel fish, and the clicking of clown triggerfish.

On February 9, we were ready for the big day, or I should say the big night. Joseph would take us far out to a place he thought was a mating site for groupers. Once a year, when the moon is full around the third week of January, they leave their holes and come together in the thousands to mate, traveling great distances to participate in their mating dances. This display, whose location is known to some fishermen, obviously leads to huge catches. It is while they are mating that they emit their best sounds. During this spawning they rub against each other, creating the stimulus that causes the females to release thousands of eggs and the males their seed. This is the point at which the groupers are talkative. We wanted to be the first in the world to record those hours of pleasure, browsing in the erotic book of nature in action.

A few days earlier, I decided to dive in the area to reconnoiter. It was at the edge of the continental shelf, with a depth of twenty-five meters. That day only a few nurse sharks sinuously glided by in search of food. In the numerous holes, caves, and crevices that made up the coral, many other species were carrying on with their daily lives, but there was not a hint of a grouper. I hoped, as was normally the case, to find the young females first on the scene, to observe their change of color before mating, their frenetic mouth-to-mouth movements, their rubbing up against the males, to see the strings of eggs streaming from their bodies, to witness the approach and withdrawal of their mating dance. But there was nothing. I also wanted to watch the rush of angelfish and kingfish on the released eggs, few of which survive to become alevins in the nearby sea grass beds, thereby continuing the cycle of life. I concluded that I must be one day too early.

Our hydrophones were in the water during the ensuing days and nights. To our good fortune, the sea was calm, and the lack of waves made our work easier. In the silent, humid evenings, alert to the slightest sound, our unoccupied vision imagined furtive movements underwater, but we heard nothing that might be of interest to us. Either the site had been deserted by our friends after the massive catches by fishermen in recent years, or as Patrick, a fisherman from Saint Barts known as "Crazy Pat," later suggested, because it had simply been abandoned for another site at a greater depth.

The last night, discouraged but not very surprised, we were rewarded by recording the distant mournful sounds of a school of whales. The long, melancholy horn calls typical of the species enabled us to detect the presence of several calves who were indicating their position in the darkness to their mothers.

The next day, probably as disappointed as we were by our failure, Joseph brought us four huge lobsters.

It was now time for us to change anchorage. We chose to set anchor on the west coast near Low Bay, facing a small palm grove overlooking one end of the lagoon, while the other end was a huge beach with dangerously steep slopes. This topography told us that a strong swell from the west must often prevail here, and we therefore anchored a good hundred meters from the beach.

We were now near the northern zone of Barbuda and not far from the capital, Codrington. The only problem was that in order to cross the lagoon, we had to drag our dinghy and its motor up the steep slope of the beach, then push it across the remaining hundred meters of sand to get it into the water. On the other hand, the whole area running from Cedar Point to Boar Point was very accessible to us.

John appeared punctually to meet us on board. Having spoken to Joseph, he already knew what we expected from him. His area was also divided into three parts: lagoons and mangroves, channels and shallows, and finally, reefs and open water. Sometimes we would follow him in our dinghy, sometimes we would go with him. We would also alternate our trips between day and night. We would begin on Monday, because Sunday was for worship. That gave us a day off.

From the top of our mast, where I found myself in the morning in order to get a general view of the surroundings, I noticed a large number of frigate birds circling above. They must have caught sight of a spot that was already familiar to them. Once we'd pushed our dinghy over the sand, we went toward the spot. In the northwest area of the lagoon, after following many winding channels lined with spiny shrubs and mangroves, devoured by mosquitoes, we finally came in sight of the frigate birds. Using our oars and as quietly as possible, we drew near an unforgettable spectacle. Through the bushes and tree branches we could admire all the males of the flock strutting about to seduce the females. To that end, each of them puffed up the sac of bright red skin beneath its hooked beak; this sac, often larger than the body of the bird, the color of which has to be as bright as possible, is their chief mating device. Try to imagine this vision of gray undergrowth against an azure background spotted with large red balloons. Immobile and potent, a male bird would wait for a female stimulated by the redness of his skin to deign to set down beside him before beginning a shrill conversation that would end with the formation of a real couple. Since that day, frigate birds have been dear to our hearts, and we will never criticize them when, swiftly flying by, they steal food out of the very beaks of other species.

The ensuing weeks were full of hard work. At night hiding in the branches of the mangroves, listening to and recording the gobies and sardes, diving during the day to locate the right species, I filled my recorder with all the sounds that seemed interesting to me. John was fascinated by our work. He was prompt to show us good spots, to take risks on our behalf, often to sail into open water despite bad weather, to help us discover the northern zone as though it were his field or farm. During these weeks, we recorded hours of sounds on our minidisks. Once again, we observed that for some fish the dialogue also took place during the day. One night, when John could not come with us on the dinghy, we once again allowed ourselves to be caught by bad weather. This time we were equipped with powerful searchlights that allowed us to make out the narrow strip of beach and to return to the boat without too much difficulty, enriched by new sounds.

The immense stretch of water near the capital provides many species with a safe place for mating, and I therefore decided to set our hydrophones there at the end of our stay. John, who was with us that day, was very surprised when I told him that the sound of bacon sizzling in a pan meant there were thousands of shrimp at the bottom of his lagoon.

Our relations with the residents of Codrington had become very warm thanks to Joseph and John. We were invited on several occasions to explore their island.

One day, the sea was stormy and a driving wind was lashing the *Prince*, which was tugging bravely on its anchors. Elvin, a friend of John's, called us on the onboard radio to invite us to take a spin in his pickup truck to the Highlands in the north. When our dinghy reached the other side of the lagoon, which that day looked like the Bay of Biscay, we were soaked. Elvin and his friend Alex invited us into their modest hut for some hot tea. The little house with cinder-block walls had two windows, with brightly colored and ill fitting shutters. Its roof was made of dented sheet metal set on a frame of scrap wood. The beaten earth yard, filled with countless objects, most of them rusted, was also used to raise tortoises. Sitting on benches around the table covered in oilcloth and looking around this little rudimentary room filled with old boxes, I recognized the extreme insecurity of our new friends' lives. An odor of dried fish and damp earth drifted through the humid air.

We followed a sandy, barely drivable track through spiny vegetation, groves of shrubs resembling broom, and huge fields in which the skinny goats seemed to have gotten lost. A few wild deer scattered as we went by. Farther on, at the side of the road, an old fisherman had set up an old wooden shed to live in and had decorated the stunted trees around it by strewing their branches with all the multicolored buoys he had probably collected on the beach. It wasn't Christmas, but it certainly looked like it, at least the colors did. When our track returned to the strip of land without vegetation at the edge of the sea, Alex parked the truck on a rocky platform overlooking a wild beach.

The Atlantic coast was open to the full force of the ocean winds. Béatrice and Simon, who were with us, had difficulty standing and walking on the rocks. We were now in sight of the cliffs, sixty meters high, overlooking the coast.

When we reached the foot of the cliffs and looked up, we had the impression of being towered over by the huge wall of a natural fortress. Halfway up, scattered among the vegetation, a large number of dark holes, which must have been cave entrances, made the place even more mysterious — mysterious enough, in any case, to make us want to explore some of them.

Gaining access to them, although dangerous, was not hugely difficult. Some cavities, immediately very spacious, could have provided comfortable shelter to a whole tribe, while others could be entered only by squeezing through a narrow passage leading to many large vaulted chambers at the back of which many other caverns high enough to stand up in seemed to stretch into the darkness.

The next day, equipped with good shoes and large flashlights that we had on board, we decided to explore some of them. By chance, sliding into one of the many passages that led from a huge chamber toward the depths of one of the caves, holding on as well as I could to the walls, on a slippery surface, and avoiding touching the fragile stalactites, I found some strange reliefs carved on one of the walls. I was just at the entry to a smaller chamber, barely illuminated by a weak light coming from the top of the vault. Before alerting my friends, I took the time to light up my discovery from every angle. There were two roughly carved portraits as large as faces. Their outlines were clearly defined, eyes and mouths probably shaped by strong manual pressure. One was more successful than the other. By using my flashlight to light them from below, I was able to see all the forms of the faces; the rock was very hard, and there were no visible traces of recent use of a cutting tool. It was clear to me that these portraits had been made by men who had lived here very long ago. When the rest of the group arrived, very excited by this discovery, I took many photographs, drew the faces as well as I could, and traced them while being careful not to touch the sculptures too closely. Everything was turned over to the local government the next day, and we told them exactly where the discovery had been made.

The islands of the Caribbean have a fascinating history. When Christopher Columbus landed, two minor civilizations shared the islands, both originally from the banks of the Orinoco. One of them, the Arawak, more technologically advanced, was systematically hunted down by the other, the Caribs, who stole their knowledge and exterminated them. It is estimated that the Arawak arrived in the region around the beginning of our era. We know very little about the population living there, no doubt in a very primitive fashion, before the Arawak arrived. A few traces have been found on Grenada, Saint Lucia, Martinique, and Saint Croix. They were an offshoot of the great population movement that began in North America around 40,000 BC. These "fisher gatherers" had chosen this spot to live in, and it is easy to see why, as the cliffs offered complete safety, a secure shelter, and proximity to the sea, a certain source of food. Their arrival is now estimated to have been during the third millennium BC. What happened between then and the arrival of the Arawak? Perhaps we contributed in a tiny way to a better understanding of these creatures whose faces are sculpted in the rock.

Bad weather had been keeping us from using our hydrophones for two days. So we used the time to visit the little town of Codrington. The main street, one kilometer long, was lined partly with huts identical to our friend's and partly with more substantial houses, solidly built and handsomely painted. One of them, in particular, was surrounded by a charming garden in which Betsy, a tame deer, contentedly wandered about munching hibiscus blossoms. A few vans, most of them in disrepair, parked in front of the doors were signs of the relative wealth of the owners of the houses. Free-running thoroughbred horses were grazing near a rough racetrack, probably getting ready for the Sunday holiday.

Numerous churches, which we had a good deal of difficulty locating because their architecture so closely resembled that of neighboring houses, were scattered around the village. Every one of them was surrounded by a little garden with alleys lined by stones painted white, representing, it seemed, one of the many tendencies of the Anglican Church. We weren't sure that we'd understood everything.

Since it was Sunday, we heard waves of music in the street, and since the whole population of the village was in church and we had nothing better to do, we went into one of them. The spectacle was barely beginning. I must beg my friends' pardon for using that word, but in my view it's not insulting, quite the contrary. The joy experienced in those moments while music and rhythmic singing elevated the soul was a pure pleasure for all of us. It must be said that all the participants were dressed in their Sunday finest, particularly the children. Hair with multicolored ribbons and slicked down with shining gel emphasized the transparency of the black skins; the well-worn but formal gray jackets on the men with callused hands, the long shapeless dresses worn by older women with fancy hats, no doubt following island fashion, the altar decorated with palms, behind which were piled up the instruments of the orchestra — all of that seemed to have been arranged by a brilliant stage director. No detail was lacking: in those expressive faces with eyes glowing mischievously, some mouths with perfect teeth contrasted with others with only three or four front teeth left. Everyone was immensely pleased to be there, and so were we.

Three full hours went by in a flash. The rich smell of perfume and perspiration blended with the incense, the hand-clapping in time to the chanting of colorful hymns, and the syncopated music provided by a discordant duo of guitar and saxophone all combined to make the time seem too short.

At the end of the sermon, surrounded at the altar by his most faithful female parishioners, the pastor lifted his arms and spoke to me.

"I see that in the back we have the pleasure of having some white people among us on this holy Sunday. May I ask your names and what brings us the joy of welcoming you today?"

"Our names are Simon, Béatrice, and Gilles, and we come from France."

"And what are you doing on our lovely island? On vacation?"

"No, Reverend, we are here on a scientific expedition."

"Very good, and what is involved?"

"We are listening to and trying to understand the language of fish."

"And what do the fish say?"

Prompted by the naïveté of his question, the wonderful simplicity of the ceremony, or the emotion of the moment, I don't know what came over me, but I answered:

"They're praying to the Lord."

After the hubbub in the congregation died down, he responded, speaking to the whole congregation:

"You see, even the fish pray to the Lord. And you only come here to pray once a week. Learn a lesson from these creatures of God. Innocent beings who never forget the Lord. That's an example to follow, and it is given to us by science."

And the pastor launched into a magnificent tirade that made his congregation both tremble and laugh. He was interrupted a few moments later by the musicians playing a rhythmic jazz melody, probably to calm him down. What an extraordinary moment that seemed to fill a need. We were all invited to share the refreshments that followed.

After that day, all the residents smiled broadly whenever we met them in the street.

Miraculously, good weather returned the next day. The ensuing days were taken up by our recordings. John took us to the coral reef at the northern end of the island.

While working at that location, I was able to perfect a method for listening in a hostile, underwater environment. I did this both by developing a system of locating buoys and by blocking out various square-patterned areas. Observations and advice from f122ishermen were of great help. I quickly learned how to detect possible mating areas, particularly for the porgies, work out their behavior patterns, and find the routes they traveled during the day to find food. I also determined the technique that they used to come together before nightfall, their resting places, and their mating habits — in short, everything for which I had not been trained during my stay at Cape Canaveral.

The northern zone, just outside the channel leading to the lagoon, between Goat Island and Billy Point, was the ideal spot for putting this method into practice: first because the area is well protected from the open ocean, and second because it is bounded on both sides by forests of mangroves, whose roots are used by fish as a mating site. The spot is also sprinkled with holes dug out by the strong tide that fills the lagoon twice a day. It was a great pleasure to work in this landscape, where nature had placed on the water's surface some vertical branches whitened by the sun that were used as perches by herons and pelicans, and had carved out the mysterious channels in the forest that concealed unknown fauna. Some evenings, just at nightfall, the reflection of the setting sun on this living mirror gave the place an unreal air; the colors were so vivid that they seemed to set the limitless space ablaze.

Rather late one evening, after the shadows had disappeared and only moonlight illuminated the silvery scene, we ventured into one of those channels with our hydrophones. Advancing through the calm night, with a wall of mangroves on either side, our fragile craft left a straight and gentle wake on the dark mirror. We were all silent, trying to detect in the darkness the right place to stop and listen to what might be going on in those dark waters. Finally settling on a spot, I put on my earphones and dropped my equipment beneath the surface. Nothing, absolutely nothing, not a sound. Suddenly I heard a huge muffled bang that I was unable to interpret. John and Simon jumped in front of me, unbalancing our craft. Absorbed by my listening, I had seen nothing. A huge animal, it seems, had come out of the water in a flash, had passed above us, and come down on the other side. Science had its limits. I bravely decided to reverse course and return on board the *Prince de Vendée*.

I had acquired a new name a few days earlier. Bungy, John, Moose, and the other fishermen now called me "Rockhind," a pleasant nickname that I accepted out of friendship, which normally is the name of a kind of yellow grouper with light red spots, known for its large eyes enabling it to see very well. My glasses must have been the reason.

The following Sunday afternoon, John invited us to the races. All the riders were getting ready in one corner of the field. A dozen men wearing multicolored riding caps were putting on T-shirts with labels for brands of Antiguan beer.

Bets were placed in a wooden shack built for the purpose. John had his favorite. The paying spectators were seated on wooden stands. Huge loudspeakers were playing Bob Marley at deafening volume. The whole village was there, even the dogs, running among the horses' hooves and spooking them. They were off. At the first notes of the song "Get Up, Stand Up," shouts of encouragement flew from the crowd. I could not see the winner because of the crowd around the finish line. John lost his bet, which I learned from the volume of curses he let loose next to me.

That evening we were dinner guests of Elwyn's family. Frank, the head of the fishermen, would be there too, which would enable us to tell him about our discoveries. Grilled lobster marinated in beer, turtle soup with beer, and beer cake, all washed down with beer — it was a great meal. The atmosphere in the dining room with lace curtains was certainly warm. Too bad John's horse hadn't won. The table that had been so carefully set now looked like a battlefield; everyone got up to dance in his own way, paying no attention to his neighbor, to Jamaican music coming steadily from an old radio. How did we cross the lagoon in the middle of the night, pulling our dinghy one hundred meters over the sand to reach the sea, and traversing the waves in order to finally get on board? This was another mystery. The next day we went back to work.

Our "chef" Béatrice, as planned, was to return to Europe. I gave her all my recordings so that she could send them by FedEx from Antigua to Professor Gilmore in Florida. We saw her off with regret at the little airport in Codrington. Her courage and simplicity had won us over. Her vanilla pudding would be sorely missed. Our time in Barbuda was also coming to an end.

The following week, we had a visitor. Dominique, our friend from Antigua, came to join us in her boat, a polyester launch with two very powerful outboard motors. Probably having left a little late, she arrived at nightfall. Since morning, a heavy swell from the west had been tossing our catamaran about. Although we were anchored at a fair distance from the beach, the backwash created by the undertow on the steep beach was now breaking fifty meters from our prow. Our situation at that point was not very comfortable, but it was too late to change our mooring because we were surrounded by coral outcroppings. We decided merely to increase the security of our two anchors until daybreak. Dominique dropped anchor a quarter cable to starboard of the *Prince*, and we went to pick her up in our dinghy. It was already almost dark when we got back on board for the dinner we had prepared for her. We were happy to see her and obviously had many things to tell her. Two hours later, before turning in for the night, she wanted to get some things that she had left on her boat. Armed with a strong light, I set out with her in our dinghy. We tried to locate her boat, which should have been very close to ours, but it had disappeared in the night. Universal panic.

I quickly brought her back on board and left again with Simon in search of her boat. I feared that the coral reef, two miles out to sea, cut through at spots by narrow passages, might have stopped it. The night was particularly dark, a heavy tropical rain was falling on our search, and visibility was zero. We spent a good three hours in this absolute darkness, in a region infested with coral outcroppings close to the surface and that we were very familiar with, searching for the boat with a white hull. Nothing, absolutely nothing. It was incredible. The boat had vanished into thin air!

Returning to our boat, furious at our failure, I immediately marked our position on the map, estimated the velocity of the east wind, calculated the approximate speed of the current at the spot, and determined the speed at which her boat was drifting and its position from hour to hour. It was one in the morning, and at three knots per hour, the boat must already be twelve miles out to sea. It was impossible for us to go looking for it in the current poor conditions. The cleat on the bridge holding the anchor chain must have broken because of the heavy seas. Dominique had left on board her bag with all her credit cards and important papers. We spent the night reassuring her that at daybreak we would return to Antigua and on the way send a radio message so that a private plane could fly over the area where we guessed her boat would be located.

Everything went as planned; fifteen miles from the coast of Antigua, we contacted the airport to report the loss of the boat and its estimated position. A plane took off thirty minutes later, and one hour later we received word over the radio that the boat had been found in the area we had predicted. All we needed now was a speedboat to tow it to Antigua. When we were again alongside the dock in Jolly Harbour, everything was practically settled. Dominique's bag had been found still on board her boat. The quick decision had saved the situation; the very heavy sea would never have allowed us to catch it ourselves. Two days later, her boat was docked facing her charming house.

That evening, with good spirits restored, and despite the expense of the rescue mission, we found ourselves in Antigua seated before a good dinner prepared for us by Dominique.

This little improvised break did us some good. We could replenish our supplies, sleep without worrying about our mooring, and rest a bit before setting out again.

Some minor damage was repaired, the fuel tanks filled, a few e-mails sent out, and we were once more on our way to Barbuda following a good east wind.

I chose the same mooring, just opposite a coconut grove. The waves had calmed. During the first day, using our diving tanks, we tried in vain to recover the lost anchor from Dominique's boat. The three days of heavy seas must have totally buried it in the sand.

All our fishermen friends were waiting for us. Worried at no longer seeing our tall mast rising above the beach, they had come to the conclusion that the bad weather had made us leave the island without saying good-bye.

Our work listening to the fish resumed, this time with the help of Elwyn. The technique of the buoys worked marvelously, and our recordings of the "clicking" of the porgies during the day confirmed our discovery that they could communicate at any time. This was completely contrary to Professor Gilmore's opinion. We could now discern three sounds of differing intensities: one, rather high-pitched, to mark out territory; another, rather short, to call females; and a third, very staccato, to indicate danger or drive off an intruder. If you recall my walk on the cliffs with my ornithologist friends in California, it seems that I was coming to be able to recognize fish by their sounds.

To verify all of this, I asked Elwyn to arrange for me to dive with my underwater camera not far from the spot where we had found a family of porgies. Since the fish were not very timid, I decided to follow the largest one to photograph it. This little game of hide-and-seek lasted for a good hour, with the fish hiding from me and with me discovering his hiding place. I always maintained a reasonable distance in order not to frighten him off. Finally, exasperated by my persistence in following him, he turned toward me with a rapid movement and emitted a clicking noise that I could hear perfectly clearly underwater, as though he were telling me to lay off.

In any event, that's what I understood him to be saying. If the pastor had been with me, he would have said that that one had a soul.

Days and nights passed quickly. We had been here for a month. We had to prepare to sail to Saint Barts and Saint Martin, where we were scheduled to give lectures.

After having accumulated a mass of recordings and notes, I decided to devote the last few days to our friends. We shared a modest meal with one, invited all of them on board, took pictures of families, took leave of the pastor, visited Frank's turtle farm, and fed John's horse. He had no doubt that his horse would win the next time. We also had to take leave of Betsy, Moose's tame deer, thank Mr. Nibb, head of the local government, and visit two pretty girls who were in love with us and wanted to give us a chaste kiss before we left.

We visited the little museum to see whether the photographs of the sculptures we had discovered were well displayed. We said our farewells to the telephone operator who had arranged our calls to the United States. We turned over to the fishery department all our detailed maps of the island, which they had very humbly requested a few weeks earlier. We shook hands with unknown parishioners who still remembered the fish that prayed to the Lord. When would we return? Would it be before Christmas, or not until next year? These were questions we were unable to answer.

The last night, it was very hard for us to leave. Crossing the lagoon with our dinghy and dragging it one final time across the beach to get to the *Prince*, our eyes were a little damp, and yet, for once, it wasn't raining that night.

The next day at dawn, we steered a course at 280 under a driving rain. Very large black clouds covered the horizon. Behind us to the east, a bright sun rose in the clear sky over Barbuda.

5

The Gannets of Dog Island

A stolen tuna — Landing on a so-called paradise — Sailing to Dog Island — A stormy mooring — Whale cemetery — The gannets, our friends — A hidden airport — The Galapagos of the Caribbean.

We had nothing but bad weather for the seventy-mile trip to the island of Saint Barts, although we had no reason to think that low-pressure fronts from the United States would continually cross our path. They brought with them not only large rain clouds, but also gusts of wind that, along with the high seas, made sailing difficult and tiring. It was only when we could see the outline of the island's mountains that the sun peeked through the bank of nimbostratus that had been streaming above us ever since we left Barbuda.

Seen from a distance, the island was very beautiful. The green slopes rising to the peaks contrasted with the flat and barely visible horizon we had left behind us when we regretfully sailed off from our long and fruitful stay on Barbuda Island.

A few miles from the coast we ran into dangerous waves five or six meters high threatening to swamp the *Prince de Vendée* at any moment. But this didn't stop us from catching, on one of the lines we were trailing, a large yellowfin tuna that must have weighed at least thirty-five pounds. It wasn't easy to get it on board.

Only when we sailed past Point Nègre, which protected us from the open ocean, did we reach calm waters. We were then able easily to make our way into the charming little port of Gustavia.

Christopher Columbus gave the island its name in honor of his brother Bartolomeo. When he discovered it in 1494, he certainly had no idea what it would become by 2001. Although they did not destroy the charm of the spot, we were surprised by the large number of extremely luxurious boats and the crowd of tourists on the docks. We were light-years away from Barbuda.

The last time I had visited was twenty years earlier. At the time the little cement roads were in good condition. There was little traffic and Saint John's Bay was completely unbuilt. Now only the church of Lorient had not changed. The marble slab to the left of the entry still indicated that it had been burned by pirates in 1647. Quail Cove, when I had been there before, had been a refuge for many painters who had come to find calm far from the capital, and there were no parking lots at Grande Saline.

The *Prince*, on the other hand, must have been satisfied,

and even happy, because he was meeting up with all his ancestors who had emigrated from Vendée to the island in 1665.

After backing our boat up to the dock and dropping a good anchor in front, I transported our tuna, which was too large for our ice chest, to the refrigerator in the restaurant across the way. I planned to present it to M. Gréaux, the local representative of the foundation, who was expecting us for the lecture I was to give for his clients and friends.

Roger, whom we had met at Antigua, and who was to join our expedition with his boat, was there to meet us. His boat, a fifty-foot Bénéteau with a navy blue hull, built in the United States, was moored between two buoys at the far end of the port.

After crossing the port in our dinghy, carefully avoiding traffic, we went to the warehouses of M. Gréaux, which were filled with rum and other liquors. This friendly man had organized a press conference in a nearby restaurant for seven the following evening. A screen, a projector, and microphones had been arranged for. In the meantime, he had set up an interview with Pierrette, a reporter on the *Journal de St-Barth*, who later became a real friend and supporter of our expedition.

That afternoon, I met Franciane Gréaux, director of the marine nature preserve, and its president J. C. Plassais, both of whom were very interested in our research and invited us to come back in 2002 to continue our listening using their boats.

The next day before the lecture, I went to pick up my tuna to present it to our host. It had disappeared into the stomachs of the customers of the restaurant. Progress had brought not only good things.

The lecture was a success and lasted very late. The audience asked many questions that I tried my best to answer. It should be said that it was rather unusual for a scientific expedition to stop here, the place being known primarily as a vacation resort.

Through a fisherman friend, I met Marc Thézé, manager of one of the most beautiful hotels on the island, the Guanahini, who also invited us to come to give a lecture to his clients in 2002.

At dawn the next morning, under a driving rain, we set sail for the island of Saint Martin, with Roger's handsome boat in our wake.

The crossing presented no particular difficulties. Only fifteen miles away, the island seen from afar, wide at the base with pointed peaks, seemed rather welcoming. I regretted that we did not have the time to stop at Fork Island, which we left to starboard, because it seemed far enough from everything to be a good listening post.

Because of our shallow draft, I decide to get to the royal marina by sailing under the Simpson Bay drawbridge in order to reach the lagoon in which a channel would lead directly to our dock. Because of his deeper draft, Roger decided to go the long way around, sailing past Basse Terre Point to reach Backwater Bay, which also led into the lagoon. Régine, who was in charge of bridges and harbors, found a mooring buoy for him just opposite ours.

The lagoon still showed signs of damage from the latest

hurricane. Many smashed wrecks were strewn on its shores. A few masts, like crosses stuck in the water, emerged from the calm surface of this ships' graveyard. We were told that the previous September thiry-five-foot waves had torn through the area. Indeed, when we approached, all the buoys for the channel providing access to the marina had disappeared. But for the moment we were in calm waters, very close to the little town of Marigot.

Excellent news was awaiting us. All our recordings had been received at Cape Canaveral. Professor Gilmore had sent a long congratulatory e-mail. He had particularly appreciated disks 3, 4, 7, and 8. "You have done a wonderful job," he wrote. You may imagine how pleased we were. The methodology that had been improvised on the spot turned out to have worked. All the risks we had taken and the dangers we had overcome during the nights of bad weather were worth it.

Florence, public-relations officer for the local division of Bacardi, told us that there would be a welcoming cocktail party on the dock right next to the boat in three days, with more than a hundred people in attendance. I made a list of friends to invite as well: Jean, René, and their families, along with Olivier and Jacques, two journalists posted here.

Since our sailor Simon was planning to leave us later when we stopped at Puerto Rico, I hired Raphael, friendly and something of a tinkerer, who would join us by plane in two months.

I prefer Saint Martin to Saint Barts. First, the open spaces give less of the impression of being on an island, and second, the division of the island into French (Saint Martin) and Dutch (Saint Maarten) sections has created a more varied landscape.

I liked the little restaurants of Red Bay and the expanse of beach in Great Cove, the charm of Pinel Isle and the turquoise water of East Bay. The docks in Marigot, lined with restaurants with exotic names, and its fruit market with multicolored stands run by Haitian merchants, always had a festive air.

We had a few days to prepare for our next sailing. The boat and all the equipment on board were rinsed in freshwater, photographs developed, fuel tanks filled, the dinghy restitched, our diving tanks refilled, our gas bottles exchanged, and finally our provisions restocked.

I also took the time to write a couple of articles that I sent to our friends in the press, along with countless e-mails to verify our meetings with the scientists who were to join us for our upcoming missions on Saint Croix and Dominica. Finally, I set out in detail our sailing schedule for the next few weeks. Roger would go with us to Virgin Gorda, the easternmost of the British Virgin Islands.

That evening, the dock was full of people. Point G, the restaurant across from us, had done things well. Nothing was lacking. There were even expedition banners hanging from the red-painted tin roof. I finally realized that the simple idea that I had had in Miami two years earlier had become a happy reality. We were among those who were lucky enough to fulfill their dreams.

On a bright morning, under a clear blue cloudless sky, we sailed from the dock. Régine opened Marigot Bridge. We headed for Anguilla. We had authorizations to obtain before we could sail to our secret island.

The flat elongated shape that eels (*anguilles*) usually have fit perfectly with our next port of call. This probably explains the name the Spanish gave it. We dropped anchor in Road Bay at the north end of the island for the night. Some wrecks of freighters that used to sail between the islands had been driven against the coastal cliff to the right of the huge beach of golden sand.

The customs and police post was facing us, barely concealed by a row of bougainvillea. Accompanied by Roger on our refurbished dinghy, we drew alongside the little wooden jetty, that sits on its stilts perpendicular to the beach. From our mooring to the edge of the beach, the turquoise water was incredibly clear. This prompted me that morning to resume my good habit of shaving on the rear deck and diving into the water. I wanted to be presentable for our dealings with the local authorities.

Contrary to our expectations, we were well received. The Saint Martin newspapers, which also circulated in Anguilla, had written about our visit. We were given an exceptional authorization to moor at Dog Island for seventy-two hours.

Flying over the area a few years earlier, I had seen this island from the airplane window. This little scrap of land isolated in the middle of the ocean had made me dream more than once. I had imagined it to be an ideal place to live for a while. Seen from above, it had a verdant appearance, urging me on to a voyage of discovery. Patches of green, a beautiful sheltered beach, a few springs, and steep cliffs had for a long time aroused my curiosity. I later learned that the island had become a nature preserve, which helped explain the interest our activity had produced in Anguilla.

Dropping anchor in the narrow bay open to the west, we disturbed some rays calmly loitering on the sandy bottom that rippled up to the beach. A little farther on, a little school of mullets were zigzagging beneath the clear emerald surface of the water. Behind us a large rock divided the opening of the bay in two. In front of us, the beach, slightly rounded and with a worryingly steep slope, was topped by thickets of spiny vegetation. To port, a rocky promontory blocked our view of another bay that we had located on the map. To starboard we glimpsed in the distance the bluish outlines of the tops of the hills of the island of Anguilla.

To reach land, we had to deal with the heavy surf crashing on the beach. Two or three cautious trips in the dinghy were enough to get us all high up on the beach. If we were properly informed, it would take us an entire day to explore the island. Once this mapping out was complete, we could then concentrate on one or two spots to film the following day.

A lagoon with brackish water, hidden by a dune, stretched between two hills covered with flowering cactus plants and mancenilliers. The thick mud on its banks provided an ideal refuge for countless mosquitoes. There was no trace of a human footstep. Farther on, when we reached the top of the cliffs overlooking the indigo sea, we were met by a surprise. A large colony of yellow-footed brown gannets was established there. Some of them were standing and chattering as though they were debating over the best spot to fish, while others, lying in their nests, were lovingly protecting their young, whose down was white, under their wings. By lying quietly on our stomachs we were able to carefully observe the birds. We had to crawl a good hundred meters to go around them and continue our journey. Farther on, a small valley separating us from another cliff farther to the east presented the first difficulties. The dense growth of spiny shrubs was very hard on our legs, but we did make it to the other side. At the top of the next cliff, a nesting place for thousands of frigate birds, we finally had a panoramic view of the northern coast of the island. From this spot we could see two more colonies of gannets. There must have been hundreds of nests on the bare ground. Below, a beach of pebbles mixed with fine sand battered by the wind was strewn with what we thought at first were the trunks of trees whitened by the saltwater that had been deposited there by the waves. We later discovered that we had been mistaken.

Continuing on our way, we were unable to reach the beach from the spot where we were. We had to follow a rocky slope, the upper part of a limestone fold, which is common in the region, to reach the western part of the island. On the way, in the lowest and greenest part, which was still full of spiny vegetation, we came to a flat space that must have been used a few years earlier as a landing strip. This field was striated with deep furrows every hundred meters that prevented it from being used. Moreover, to one side, hidden in the undergrowth, the dismantled wreck of a small airplane was quietly rusting away. The age of drug trafficking also had its monuments.

Late in the evening, our legs covered in blood, we returned to our mooring. A good swim would, we thought, soothe our wounds and abolish our fatigue. It was at the very moment when the sun was about to set beneath the salty golden surface of the sea.

The following day's schedule was clear. We would return to film the gannets, taking every possible precaution, and we would explore the pebble beach.

Roger's boat had a hard night. The swell from the north, which had grown heavier late in the evening and came straight into the bay, had half capsized his boat. He said that this was the most dangerous anchorage he'd ever seen. But on board the two-hulled *Prince*, we had barely felt the waves.

Wearing long pants and good shoes, and carrying all our photographic equipment, we set out the next morning at dawn. At the top of the first cliff, we found a crevice deep enough to observe the gannets without being seen. We spent the whole day on our stomachs trying to understand their comings and goings.

They never left their young alone. Mothers and fathers alternated in searching for food. Once the food had been digested by the adult, the chick instinctively found the spot next to its parent's beak where a peck would trigger the process of regurgitation and discharge, providing it with food. Other less responsible gannets invented games in which they squabbled, jumped, and suddenly flew off above the cliff. When frigate birds came a little too close, the mothers, while covering their young with their wings, straightened their necks and beaks and made staccato clicking noises to tell the intruder to leave immediately, that flight was the best possible choice. The community was perfectly organized. Those who seemed the oldest and most experienced constantly flew over the colony. Their spectacular gliding flight, without apparent movement, following the air currents flowing up and down the cliffs, must have had a protective effect for some and a troubling aspect for others.

The principal purpose of these flights was always the same, the search for food. In sign language, I told Simon that I wanted to get into the water with my flippers and mask to be able to observe them more closely when they made their spectacular dives. He was to silently come closer to the cliff to see whether I would be in any danger. We did not know whether, when I was on the surface, the birds, diving like projectiles, would take me for a target. I did not want to have a bird's beak planted between my shoulders. The rest of the crew would remain hidden in the undergrowth.

Having descended the easiest slope of the cliff, I was now at the water's edge. The slippery rocks that had been smoothed by the sea were not very welcoming, and I found the very high rock face rather troubling. I had the impression that I was directly beneath a huge aircraft carrier whose uncontrolled missiles would land at any moment on my poor defenseless body. Perhaps I should have been less fearful, but they seemed menacing.

Once in the water, floating on the surface, immobile, I had to be patient. It was only after twenty minutes that I was finally able to witness the first dive of a gannet. It had broken through the surface in a flash, leaving behind a streak of silvery foam. It must have been a few meters away from me. Using its wings as oars and pushing with its webbed feet, it pursued a fish that it finally swallowed greedily by sharply throwing its head back. With a slight twist of its body, it quickly rose to the surface and, with a powerful beating of wings, returned to flight. I witnessed five or six of these dives. Each time, I wondered how they could see so well beneath the surface and, when they dove down five or six meters, what effects the pressure had on their bodies. Each time they came back up, the frigate birds, who were clever but afraid to get their wings wet, tried to steal the concealed prey. Astonished by the spectacle, I climbed the steep slopes of the cliff almost unaware of what I was doing.

During the all too brief time we were there, we never saw one gannet trying to steal the nest of another. We observed the unconditional love the mothers displayed to protect their young. Here there was only one chick in each nest. A little farther on, beside one of the nests occupied by a mother and her chick, we had noticed a little corpse. I concluded that this was an example of the harsh natural selection of the species. The fittest survived, as usual. We had the impression that the chicks were always famished. The parents constantly took turns feeding them, putting off to less stressful days the pleasure they had in flying with no apparent purpose.

We never had the impression that we had disturbed their daily routine. We were well hidden, but they were probably instinctively aware of our presence. Nevertheless, the behavior of the gannets did not seem to have been altered. There are nine species of gannet, whose names vary as a function of their habitat or the color of their feet. They always live in huge colonies and feed exclusively on fish. The nest, often rudimentary, is made of little twigs roughly piled together. They incubate their eggs for forty days, keeping them beneath their webbed feet. We left them regretfully by crawling backward. It was time to allow these flying machines to continue their aerial ballet in the azure sky.

The experience, marked by great tenderness, will stay with us for a long time. We all hoped that other less scrupulous people would not come to trouble their existence. Our photographs were a guarantee that we would have pleasant moments that winter when we went through our documents.

The next day, taking an easier path, but one still covered by brush, we finally gained the beach that we had seen two days earlier. It took us nearly three hours to reach the spot, even though it was very close to the mooring. Cut off by large overhanging rocks, the beach, half covered with pebbles, was swept by huge waves. The wind, which had freshened that day, gave the raging sea the appearance of a field bristling with millions of wandering sheep. What we had taken from a distance for uprooted trees deposited by the waves were in fact the bones of whales. There were dozens of them. Vertebrae and jawbones were strewn over the entire surface. As well as we could, we tried to reconstitute the body of one of them, without success, since the sea had probably taken back what it had once cast off. Was this an old whale graveyard, or merely a few whitened carcasses that had died by accident? We'll never know.

The sea's ill humor, for once, had prevented us from listening to our friends the fish.

Dog Island had revealed to us only a tiny part of its secrets. This little "Galapagos" of the Caribbean had so stimulated our imagination that we will have to return someday.

Well before dawn the next day, we left our rolling mooring to sail for the Virgin Islands. Before the sun sank too low, we absolutely had to get through the tricky channel between Necker Island and the little island of Eustatia, both north of Virgin Gorda.

For once, the wind was with us. We had seventy miles to go before reaching our next mooring in Gorda Sound. The sea was fine that day, our morale was high, and so was the sun.

6

Rose, Anegada, and Our Studio

A well-deserved stop — A boat on the reef — The forgotten island of Anegada — The story of Rose and Eddy — The treasure of the wrecked ships — Lost in the midst of corals — Fear and disappointment — The pleasures of the studio — The richness of the fauna — The groupers speak to us — Arrival at the Baths — A perfect island.

We neared the coast of the Virgin Islands well ahead of schedule. Roger's boat, which we had outpaced by unfurling our spinnaker when we began the crossing, was three miles behind us. When we approached the narrow passage leading to our mooring, we saw in the distance a boat that had been wrecked on the coral reef bordering Necker Island. This boat, about thirteen meters long, lying on its side, and probably with its hull ripped open, looked like a mortally wounded animal. Fortunately for it, the sea was calm and only gentle waves lapped against its only visible side. Its keel was half hanging over the reef and the rigging was hanging in disorder from the mast down its starboard side. As we got closer and looked at it through binoculars, we recognized it. We had met its crew when we put into Saint Martin.

With no hesitation, we decided to sail toward it. When we were at a reasonable distance we launched our dinghy and discovered that the boat had been abandoned, probably the day before. There was no one on board. One hour later, when we reached our mooring, we learned that the crew was safe and sound. The disaster had happened the day before at nightfall. Because of poor visibility, the captain had underestimated the distance separating him from Necker Island, relying too heavily on his GPS system. We had clearly been right to leave early in order to arrive well before sunset.

This event, which might have been even more tragic, put a temporary end to the family's travels. The boat was refloated a few days later with the help of a waterborne crane and towed to the port of Virgin Gorda for all the necessary repairs.

Our mooring, sheltered by superb mountains, was perfect. If I had been a pirate, in 1650 for example, this is the spot I would have chosen to harbor my fleet and set up camp. In 2001, it was to be our home base for the ensuing weeks of research and the point of departure for our future meetings with fishermen.

The next morning, we looked at the detailed map of the islands in order to find the best spot. We decided that the bay on the west of the island of Anegada, between Setting and Pomato points, would be our next port of call. Roger had to go back to Miami on business.

I took a boat and a taxi to accomplish the formalities for

our entry and clearance at the marina on the other side of the island, in Spanish Town.

Meanwhile, barely ten miles from us, Rose was driving her little van to pick her daughter up at the school next to the bakery in the center of the little town of Settlement, the capital of Anegada. Rose's story is truly incredible.

Rose was born in Haiti of an unknown father and raised by her mother, Jacinthe. They lived in a little hut in the woods in the northern part of the Artibonite Plain, close to the town of Gonaïves. Her mother's sole source of income was a simple stall from which she sold charcoal that she made herself by gathering the few shrubs that were still growing at the foot of the nearby mountains. Even then, the island was already suffering from deforestation. When a neighbor was able to drive her to the market in Port-au-Prince, she could always return with a few gourds, enabling her to raise her daughter. Rose, a little mulatta with large black eyes, could not only go to school but also wear pretty little dresses. She particularly liked a rose-colored one that she wore every day. Jacinthe, no doubt worn out by hard work, illness, and the lack of medical care, died when Rose was six. The old schoolteacher, who knew the family a little, informed a far-off uncle living in Cap Haitien of the sad event. He came to get her and take care of her, and Rose went with him to Cap Haitien. Her uncle, Phélimé, was a painter. His paintings displayed some talent, but they sold poorly because there were few tourists in the northern part of the island. Although she was left pretty much to her own devices, Rose continued in school until the age of fourteen. In the meantime, she helped out the neighbors by gathering wood for them and helping an old fisherman sell his fish in the market. She didn't earn a fortune, but over the years accumulated a little nest egg that she cautiously hid under her mattress. For Rose had a plan. She had now become a pretty young woman, a real charmer. She realized, even though her uncle was always kind to her, that her future lay elsewhere. A few years later, she had the wild idea of leaving the island. Thanks to her good relations with the fishermen, she was informed that a boat was clandestinely taking on passengers for the United States.

The crossing to the coast of Florida was dreadful. Some of the emigrants were swept overboard in the first storm. She herself, seasick throughout the crossing, never understood how, when they came offshore one of the islands south of Miami, she had the strength to swim to the beach. Miraculously avoiding the American Coast Guard, she was taken in by a couple of fishermen of Haitian origin. She stayed with them for two weeks until she could work out her new plans. With their help, she was able to recover her strength, get some clothes, and find a way to get to New York by bus. Fortunately for her, the few dollars that she had managed to hide in her clothing had survived her time in the water.

Rose remembered the name of her uncle's brother, Aimé Valentin, a New York taxi driver, who lived on 39th Street in Brooklyn. That was all she knew. When she knocked on the door of the house with pretty flowers in the windows, she was penniless. The woman who opened the door was Aimé's wife Marie, who came from Jacmel, a village in the south of Haiti. After recounting her journey, Rose was warmly invited to stay with them. During the next two years, she worked, illegally of course, for a neighbor who owned a dry cleaner's not far from Aimé's house, on the other side of Green-Wood Cemetery. She got rather used to this new life. She learned English and adored her new family. On Sundays, Aimé took them to see Manhattan in his yellow taxi. The first time she crossed the Brooklyn Bridge with him, she remembered it well, she glimpsed the Statue of Liberty to her left. Marie, who had no children, spoiled her as much as she could, using the little she had to buy a few dresses that Rose always wore with pleasure and gratitude.

On Christmas of the second year, Aimé and his wife were visited by Eddy, a young cousin from the island of Tortola, a mechanic who had come to New York to buy spare parts for automobiles. Rose, who was still very shy and unwilling to say much about this meeting, lowered her eyes and avoided going into details. I drew the conclusion that it was love at first sight. She joined him a few months later in Tortola, I don't know exactly how. Two months later they were married. The next year they had a daughter whom they named Jacinthe in memory of Rose's mother. Soon thereafter, Eddy's great-uncle died, leaving them his house and a little restaurant on the beach on the western side of Anegada, where they decided to settle. Eddy became a fisherman, and Rose, after some renovations, took charge of the restaurant.

I've told this story not in order to deviate from recounting our expedition — quite the contrary. It's impossible to make friends without learning the story of their lives. Rose is not the only one to have had this experience, but she had better luck than many others. During our travels around the islands, we encountered Cubans fleeing their island for the United States, Haitians landing on the beaches of Saint

Martin, natives of Dominica trying to survive on Guadeloupe or elsewhere, men and women from Saint Vincent looking for work on Martinique, women from Grenada working as waitresses on Saint Barts, young men from Nevis who had become expatriates on Puerto Rico. They were all displaced persons in search of a better life. Countless denizens of the islands had come to Saint Thomas in search of fortune. The adventures of the first inhabitants, the Arawak, fleeing their territory in search of a safer place, seemed to have set the tone for present-day migrations. Blacks, brutally imported from Africa, were also looking for refuge.

When we reached Anegada, after going through a narrow passage in the middle of many coral outcroppings protecting the island, we decided to drop anchor on a sandy bottom in order to secure a quiet mooring for the night. When we landed with our dinghy on the nearest beach, we were greeted by a pretty young woman, who had watched the arrival in her waters of our two large white hulls topped by a wide enclosed deck with large plate-glass windows. We always flew from the jib the expedition flag along with the colors of the country in which we happened to find ourselves.

"Welcome to Anegada. From a distance, your boat looks like it's frowning," she said.

This perfectly accurate observation made us laugh. Indeed, our cockpit, because of the shape of its openings, seemed from a distance to be frowning. She proceeded to sing the praises of her restaurant. That evening they would be serving lobster grilled on a wood fire.

"Thank you for your welcome. We will be three for dinner at about eight. Is that all right?"

"Okay, eight o'clock, for four, right?"

"No, three."

"Just ask for me; my name is Rose."

That evening, after an excellent meal just off the beach, under the coconut palms and royal poincianas surrounding the restaurant, we met Rose's husband Eddy. Having become a skilled fisherman, he obviously supplied the restaurant with lobsters. Cleverly, he exported his lobsters on the ferry connecting the islands to Tortola and restaurants on the other Virgin Islands. His boat, a solid laminated plastic dinghy with a shallow draft, was topped by a small wooden cabin, which protected him from the heavy squalls that were frequent in the area. It was moored at the wooden jetty just opposite the restaurant.

Although Eddy said that his specialty was lobster, we later learned that from time to time he used nets to catch fish. His method was to use lobster pots. He had, he told us, a good thirty, which he checked every day. His only problem, he went on, was that they were set rather far off, near Horseshoe Reef, six miles to the south.

He was intrigued by our work. He had heard about the sounds made by some fish, but he didn't see the point of the research. When I described what we had accomplished with the help of our friends in Barbuda, he became more interested. Once he understood that our studies might one day make possible a better understanding of his world and hence more protection for the natural environment, he was our man.

Anegada has the same characteristics as Barbuda. It is a low-lying island protected by numerous reefs. In its center, a lagoon lined with mangroves is home not to frigate birds but to pink flamingos. Sandy tracks cross the island from one end to the other. The capital, Settlement, is smaller than Barbuda's, but has the same kind of shops. There is also a firehouse and a hospital, but we saw only one church. It should also be said that Anegada is only about one-fifth the size of Barbuda and has only one thousand inhabitants. Anegada, it is said, is the island of freedom, with no crime and no theft. It is an old-fashioned and quiet spot full of island charm.

The east side of the island, which is exposed to the winds and extended to the south by banks of coral, has a very bad reputation. Many ships have been wrecked on the reefs here, and it is estimated that there are more than three hundred of them lying on the bottom. One, the *San Ignacio*, a Spanish boat belonging to the Caracas Company, sank on the reef in 1742 with a cargo of gold and silver. It had taken on its cargo in Cuba and was sailing for Spain. Another, *La Victoria*, broke up on the reefs with more than ten million dollars' worth of gold and silver in its hold in 1738, as did the *Nuestra Señora de Lorento* in 1730. More recently, the *Rocas* was wrecked here in 1929. It was carrying bones intended to be made into phosphate. Many underwater expeditions had been undertaken in the past to recover the rumored treasures. These days, authorization to conduct explorations is much more difficult to obtain.

"That's where we're going to go," said Eddy.

"Agreed for Horseshoe Reef, but are there good moorings?"

"Don't worry, I know the place."

The treasures that we were after were in the possession of the fish. The joy in discovering their language had much more value than all the gold ingots in the world, although not everyone on board would agree. In any event, we were not equipped for hunting for treasure. But some of us began to dream, including me.

Early the next day, Eddy showed us the way in his boat. Two hours at low speed brought us to our anchorage. From the top of the mast, Simon had the task of carefully monitoring our course in case an unexpected emergency forced us to leave the spot in a hurry. As for the thousands of coral outcroppings that we crossed and sailed around, we had to have complete confidence in our friend Eddy. The *Prince* and I were both on tenterhooks.

Once we had arrived and were firmly anchored, there was nothing to regret; we were in the most beautiful and probably the most dangerous spot in the Caribbean. The water around us was extraordinarily transparent, and the huge coral reef spread before us, on which gentle waves broke like tiny white teeth emerging from the sea, protected us from the open ocean. There was not a ripple on the surface, and yet we were in the middle of the sea. We could barely discern the outline of Anegada to port, while in the distance to starboard the high blue mountains of Virgin Gorda plunged into the sea. Our new friend Eddy really did know the good spots.

We were fortunate to have good weather for five days, because this apparently perfect mooring had serious disadvantages in case of high seas from the west. We immediately set to work. We could take our pick of suitable places for recording. The results were excellent, because this was a meeting place for many species. The second night, we were even able to record sounds from on board. The transparency of the water enabled us several times to identify and name the fish emitting sounds without our having to dive. This would make our friends in Cape Canaveral jealous. Eddy came to see us for lunch every day.

Two mornings in a row, he took me with my diving equipment in his boat beyond the reef so that I could inspect the condition of the coral. Although it continued to provide protection, the condition of the coral was disastrous, as it was on Palaster Reef in Barbuda. The reef was nothing but a heap of limestone on the side exposed to the winds. Several sharks, fortunately solitary, were swimming not far away just beneath the surface. They paid no attention to us, and their tranquil appearance and the sinuous movement of their bodies hinted at no animosity. Even so, we didn't take our eyes off them. Eddy, who was with me underwater, was as surprised as I was by the damage inflicted by the most recent hurricanes. In contrast, when we returned to his boat in the area protected by the reef, there was an explosion of life. The coral flourished with its countless colors and its hundreds of shapes, and schools of fish large and small were engaged in their favorite pursuit, the search for food. A symphony of colored movement filled the space. Three-colored angelfish, fire parrotfish, yellow-tailed dragonets, and rainbow labres streaked the blue background with color. Eddy pointed to two large umbrella acropora surrounded by fire coral under which was living a colony of spiny lobsters.

Back on board, Eddy explained to me that since tourist boats had sometimes ventured here, some lobster beds had been depleted. The madrepores sheltering them had suddenly become deserted. Elwyn in Barbuda had already told me something similar. I had him listen with the earphones to the subtle rustling sound lobsters made when they wanted to indicate their presence or mark out their territory. Once in Antigua, accompanied by an underwater fisherman, I had recorded from the surface the macabre cry, probably a fearful death rattle, that they make when they are speared. The cry, which I heard as lugubrious, probably had the effect of keeping other lobsters away from the spot for a time. It must have been a sound understood by the species, a specific screech of a different frequency warning the entire colony that it was imperative for them to leave immediately for a safer place. This was only an assumption, but the facts demonstrated that this is really what happened. Considering that sound underwater travels five times as fast as in the air, it is easy to imagine the immediate consequences of spearing a lobster. This is why, on Barbuda as on Anegada, lobsters are caught without violence. Some are caught by diving with wire lassos attached to a stick, others simply imprisoned in lobster pots. It would be a good idea to stop spearing lobsters.

Indications in the sky of bad weather and the daily morning weather report from David Jones in Tortola of the approach of winds from the west meant that we had to leave the mooring. Eddy invited us to spend the next few days in safe harbor in front of Rose's restaurant, which enabled us to get to know them better. Eddy, Rose, and Jacinthe showed us around their island, an experience we are not likely to forget.

When we returned to the center of the magnificent site of Gorda Sound, a huge bay surrounded by verdant mountains, we were invited to tie up at the dock belonging to the Bitter End. This luxurious hotel provides passing yachtsmen with all imaginable supplies. Its numerous private bungalows standing against the hill overlooking the hotel contain attractive and comfortable rooms furnished in West Indian colonial style. Each one has its own pine staircase, concealed by rows of hibiscus. Bitter End is a veritable little village, with shops selling souvenirs, photography equipment, food, and fashion, and its docks shaded by frangipani and oleander were swarming with sailors and American tourists.

Sandra, the manager, whose boyfriend was an underwater photographer, offered us hospitality. We were provided with abundant freshwater and access to her computer. Simon, a great fan of regattas, participated in two or three sailboat races organized by the hotel, which frequently took place in the bay. He won all of them, so the following day we proudly wore the T-shirts that he had won, emblazoned with the colors of the Yacht Club.

Sandra had organized two lectures, one exclusively for the children of the village and her clients, the other only for adults, she said.

Our experience in this area had come close to making us professionals. With pleasure, and armed with photographs and film clips, I delivered my little speech, interrupted by many questions. While the dialogue among fish always interested everyone, here it was the movement of the North American tectonic plate along the Caribbean plate that provoked the most interest. In fact, Virgin Gorda, like the other Virgin Islands, is just at the juncture of the two plates. Everyone wanted to know the consequences of this fact. The spot has a tendency to change shape because of friction at great depth, with the resulting pressure engendering the minor earthquakes that take place in the region. It is clear that a trench is growing between the islands.

We took advantage of these moments of calm to explore the coral reefs surrounding the place, particularly the reef located between the island of Eustatia and Pajaros Point. This reef, containing several passages that might be used as shortcuts into the harbor, was in the same dilapidated state as Horseshoe Reef. For the three-mile length on its east side, you could see only heaps of branches of broken coral. Entire trunks of staghorn coral were lying among the heads of coral swells that had turned gray and the wide fans of fallen gorgonia that no longer protected the little lionfish, which were so fond of calm waters, from the currents. What had become of the cycle of reproduction of the corals in this place where they normally, in a swarm of life, release thousands of already fertilized larvae to ensure the continuity of the species? The dead polyps stripped of defense left space for sponges and other organisms, accelerating their destruction. Only a few green turtles paddled calmly above this pile of limestone. Although basically herbivores, they were no doubt looking for a few jellyfish to spice up their meal. The keen-eyed moray eels, on the other hand, taking advantage of the muddle, hidden in the countless crevices, lay in wait for any sign of life passing within reach. Fortunately, these reefs on their sheltered sides, like the others, still bloomed with life, giving hope for a revival of the whole through their fragile process of reproduction.

During my car trip to carry out official formalities, high in the mountains that give this island its charm, I had noticed from the winding rode overlooking the sea another bay not accessible by land. This bay, protected from the open ocean by a reef with a white line of spray and nestled in the arms of sloping mountains, had on its right side a sort of jade green cove partially protected from the winds by a spit of land covered with mangrove trees. Looking at the map, I found that the only way to reach it was to edge your way through a narrow passage to the right of the bay seen from the ocean side. To do that, you had to go around Pajaros Point and then sail east along the peninsula to approach the passage, the width of which I did not know.

We had to leave the comfort of Bitter End and its docks to continue on our way. Our destination was the "studio," the name we gave to the bay.

The one channel, the only access from the sea into the bay, was narrower than I had thought. The strong swell from the west made crossing it tricky, because there were only ten feet on each side of the *Prince* to slip through the coral pincers. I understood why the place had not tempted other boats besides ours. But the risk was worth the trouble. We were now encircled by steep mountains covered with spiny shrubs. The two meters of depth, enough for our draft, allowed us to reach the cove, which was partially protected by mangroves. Once anchored, well sheltered from the winds in the center of this isolated spot, protected on all sides and only half a cable from the coral reef, we came to think that we had arrived in a paradise for listening. And indeed, it was a paradise that fulfilled all its promises.

For these days and nights, our dinghy was the ideal means for working. Not a ripple, not the tiniest wave troubled the quality of our recordings. We were alone in the

In the middle of the "studio," a deserted bay on Virgin Gorda, in the British Virgin Islands. On behalf of NASA-Dynamac, Gilles Fonteneau is listening to fish, wearing headphones and holding an omnidirectional hydrophone. In the background, the catamaran *Prince de Vendée*, moored in the lagoon surrounded by mangroves.

Installation of devices for detecting the movement of tectonic plates. No human power can prevent earthquakes and the tsunamis they provoke. In the near future, better understanding of the movements of the earth's crust will make it possible to warn the population of the imminence of a tsunami.

Photographing a colony of brown gannets on Dog Island, the Galapagos of the West Indies.

The author locating a yellowtail family before dropping the hydrophone in order to record their noise.

At thirty meters' depth off Las Aves, the strange ballet of
two bluestriped grunts (*haemulon sciurus*):
a tender kiss that turns into a battle.

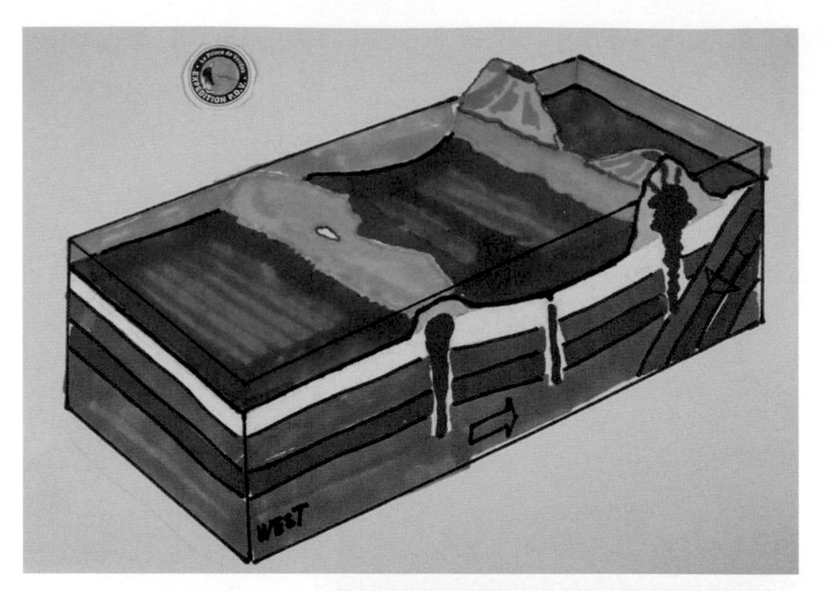

Cross section of the Caribbean tectonic plate showing the dry magma chamber beneath the arc of Las Aves.

In the mangroves in the north of Barbuda.

The strange spider with cement legs and a body of steel overlooks the island of Las Aves (300 meters long by 30 meters wide). At the foot of this fortress built by the Venezuelan army, sea turtles mate.

Map of the Caribbean on which the island of Las Aves appears for the first time in 1639. (*Courtesy of the Bibliothèque Nationale*)

The island of Las Aves seen from the air, an isolated but well defended island in the midst of the Caribbean Sea.

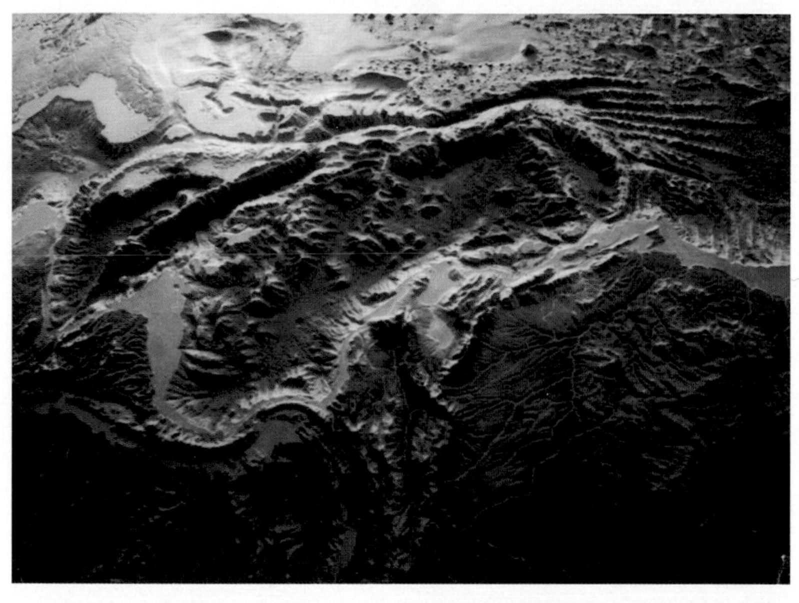

This looks like a satellite image of the Caribbean plate but is actually a photograph of a huge globe at Miami University that shows all the earth's tectonic plates.

world in this peaceful setting that seemed to have been devised especially for us. The verdant slopes, which seemed to have been delicately painted by imaginary brushes steeped in a palette made up of the most subtle blends of colors, completed this enchanted landscape devoid of any sign of animal life. Conversely, at their foot, beneath the limpid surface of the transparent mirror with its countless flashes of blue, creatures swarmed in each of the nooks and crannies of the multicolored madrepores.

These underwater depths were populated by many nurse sharks which, with their heads buried in a crevice and their appetites probably satisfied, seemed for the moment unaware of any danger. The spiny lobsters lurking under a slab, spying out the slightest movement with their large eyes and oscillating antennae, were prepared with a swift swing of the tail to scamper off to another hiding place they had already located.

This long reef also had not escaped the destructive waves. In the course of many conversations, Eddy had told us that thirty- to forty-foot waves had swept the coasts during the last hurricane. But there was no trace of disease on the corals. Not the slightest sign of moss, of white bands at the base of the trunks, of ill-defined black spots, or of infestation was visible on their surface. The salinity of the water was normal everywhere: 36 parts per 1,000. Thanks to the calcium in seawater, by synthesis of the microscopic algae known as zooxanthellae found in their tissue, the corals construct their skeleton. Water at a temperature between 20 and 28 degrees centigrade — which was the case here — fosters this transfer. Nor was there any trace in the environs of sediment that, by blocking the photosynthesis of the algae, would prevent the development of the limestone skeleton.

During the first ten months of our journey around the Caribbean islands, I calculated that we sounded out only three hundred kilometers of reefs out of the two thousand around the islands. Obviously, this was not enough to form an opinion, but I learned when I returned to Miami the following year that another group had at the same time been exploring the thousand kilometers of reefs around Cuba. Our two studies could then enable a real diagnosis to be made. Nevertheless, when Richard, a biologist from the marine center of the University of the Virgin Islands, paid us a visit when we got to Saint Thomas, I learned of the recent discovery of a virus that had appeared in the islands in the late 1980s, which is thought to be linked to domestic detergents dumped in the ocean. This long and difficult research was still going on, and samples were still being collected.

Our work followed its normal course. The most interesting moment came when we discovered by chance at the bottom of the lagoon an old abandoned, or perhaps, lost lobster pot. Swimming quietly inside this restricted space were two black and white striped groupers each a good twenty-five centimeters long. These Nassau groupers seemed completely at ease. Their behavior displayed no fear, no doubt, because their unusual position protected them from possible predators. During the ensuing week, we marked the location of the lobster pot on the surface with a buoy, and the spot became a veritable listening laboratory. We set a fairly large mirror facing the pot in order to register their sonic reactions. Their reflections in the mirror, which the groupers took for intruders trying to invade their territory, provoked some muffled and staccato sounds. When the mirror was removed, there was no more sound. Every evening, just before nightfall, it seemed to us that a brief dialogue took place between them. For two or three minutes, a series of sharper sounds pierced the silence of the depths. Not drawing hasty conclusions, which it was not up to us to do, we could observe first of all that they were able to communicate during the day, which was contrary to what we had been told, and secondly that fear of the dark, or at least a lack of visibility, provoked a sonic reaction from them. Conversely, during the night, it was not possible to hear the slightest noise. Spending all that time observing them in their surroundings, even though the conditions were not ideal, I had the distinct impression that the groupers were looking, without much hope, for a crevice in the lobster pot where they would be more comfortable. Most of the time they tried to huddle among the stones, unfortunately too small for them, that served as ballast for the pot. It was time to liberate them from this prison, which we had not built for them. I had some difficulty, underwater and without frightening them, in breaking two or three wicker strips in order to grant them their freedom. It was not until an hour later, realizing or not that there was a possible exit, that they quietly escaped and disappeared into the blue. For a time they had been the stars of our studio. Their photographs, taken during the experiment, would no doubt never appear on the cover of a movie magazine, but it hardly mattered because, unaware of their fate, they were once again part of the great cycle of life.

All the experience we had gained from our work on

Barbuda and Antigua was put into practice to collect as many sounds as possible. This was all the easier here because there were no waves and no real danger of getting lost at night. Our *Prince de Vendée*, securely anchored and fully sheltered, was close enough for us to return to it at any time. For the first time, we had the luxury of being able to forget an indispensable tool on board without that causing a real problem. We were not worried, as we had been in the past, about leaving the boat unmanned, without protection and at the mercy of the shifting winds. Here we could concentrate entirely on our search for sounds. We amassed on our minidisks a large number of new sounds. In order to put into practice what we knew about the possible reactions of lobsters when they are killed brutally, I decided to catch one with a lasso. Very quietly, I slid without difficulty beneath a slab where a dozen of them were hiding, cautiously slipped my lasso over the tail of one of them and brought it to the surface. The next day, I returned to the same spot to record their typical sounds. There was no problem, the little colony was still there. They were chattering away the same way as they had been two days before. Diving again to make sure that I was not mistaken, I saw them in front of me, comfortably settled in their hole. They had not moved from the spot. Elwyn, our fisherman friend on Barbuda, was perfectly right. Only a cry emitted after a violent wound would cause them to change their habitat.

At the end of our stay in the "studio," we were able to record a school of blue parrotfish that made a sharp staccato noise with their beaks scratching the coral to feed on the young polyp buds. In no other place did we have the feeling that a reef was an inestimable source of life for thousands of species, and that if human activity were to cause them to disappear it would be a catastrophe with incalculable consequences for our own survival.

It was with these thoughts that we passed again through the narrow exit, leaving behind us an area of undersea life that would continue to shelter for a time the few animals that we had met and tried to understand, if only slightly.

We were always happy to return to the open sea. The excitement of departure must have been a drug like any other: the unknown that tomorrow might bring, the intoxication of the moment when, raising the anchor to release the boat, you have the impression of sailing toward a new life. And then there was the discovery of new sites, different adjustments that we would have to make to approach them without too much difficulty. Perhaps there were also childhood memories, when we were pirates in attics, surging from our unconscious to make us believe again that we had a broad-brimmed black hat with colorful feathers and pistols at our belt. In any case, this is what I felt when early one morning we left this incredible inlet. And yet our next stop was not very far off.

Traveling south along the east coast of Virgin Gorda, above which large arid mountains loomed, we were now headed toward the passage dividing it from the little island of Fallen Jerusalem. This is a rocky promontory forty-three meters high given that name, it appears, because of its resemblance to that city after it was destroyed by the Romans. The middle of the passage, encumbered by numerous rocks close to the surface, was the last difficulty before reaching our mooring in front of the site of the Baths. We had come to a very touristy part of the island. There were twenty-five boats anchored in front of this lunar landscape into which were squeezed yellow sand beaches. The huge granite rocks were the result of underground pressure that pushed them up when the Caribbean plate met the North American plate ten million years ago, during the Miocene epoch. The same phenomenon, with a different origin, can be seen south of La Digue, one of the Seychelles Islands. But for the moment we all had only one idea in mind, to relax and amble under the shady tunnels formed by the huge granite masses, and then swim in the warm turquoise water without worrying for once whether we would find a good spot for listening.

We had time to kill before our appointment in Saint Thomas. It was mid-April, six months since sailing from La Rochelle. We had already covered more than five thousand miles, recorded more than forty hours of dialogue from our friends, visited and sounded fifteen reefs; it was time for a pause. Behind us stretched an interior sea surrounded by a thousand different islands.

"A thousand?" Simon asked in a worried voice.

"Yes, a thousand, that's what one of Columbus's sailors said when they were discovered. So why should I make it up? I just read it in a tourist pamphlet."

The reality, however, was quite different.

7

Fire, the Sword, and the Ever Blue Sea

The companions of Columbus — La Trompeuse — *A revolt — Our guests arrive — Aquariums in the middle of the ocean — A discovery in the forest — The anchorages of Jost Van Dyke — The bedridden captain — The hidden art of time passing — Saint John — Stopping for press and television — The new researchers — A welcoming bay — Sailing to the Spanish Virgin Islands.*

Christopher Columbus returned from his first voyage on March 4, 1493, carrying Indians, gold, and exotic birds to his meeting in Barcelona with the king, who showered him with glory and silver. Barely six months later, he was already hastily preparing his second voyage, on which he sailed from Cadiz in fair weather on September 25. He set out with a huge flotilla of fourteen caravels and three transport vessels, fifteen hundred men, soldiers, artisans, an astronomer, seeds, and cattle. Columbus also recorded that he was carrying lambs, male and female donkeys, raisins, leather shoes, cloth, two hundred suits of armor, one hundred crossbows, and ammunition. Among his crew were Alonso de Hojeda, age twenty-three, in command of one of the caravels, and Juan de la Cosa, a mapmaker. During the feverish preparations for the voyage, Columbus, who was simultaneously the object of admiration, jealousy, and hatred, constantly came into conflict with the king's private secretary, Juan de Soria, archdeacon of Seville and financial manager of the enterprise, who carefully monitored the expenses.

The principal aim of the new expedition was to recover as quickly as possible the forty men he had left behind on the island of Santo Domingo, the fifteenth-century equivalent of a space station. He was to collect as much gold as possible but also to discover new territories. Columbus had just turned forty-six. Thirty years earlier, Henry the Navigator had created the maritime school in Sagres, and four years after that Vasco da Gama set out for India around the Cape of Good Hope that Bartholomeu Dias had discovered six years earlier.

After landing on Guadeloupe, where he encountered Carib Indians, driven by southeast winds, Columbus headed back north, sailing by the islands of Montserrat, Nevis, Saint Kitts, Saint Eustatius Saba (from which he could see Saint Barts and Saint Martin in the distance), finally arriving at Saint Croix, the first of the Virgin Islands. He dropped anchor there on the morning of November 13, 1493, less than two months after sailing from Cadiz.

The log of Columbus's second voyage was lost and never recovered. Only from the stories told by the crew when they returned to Spain have we been able to learn the details of what happened when they landed on Saint Croix. Men who had sailed with Columbus, Michel de Cuneo and Guillermo Como, said that the weather was very bad, that the island was densely populated and cultivated, that the Indians fled at their approach, that they took four of the Indians captive, that others were angered and shot poisoned arrows, one of which mortally wounded one of the sailors, that they captured a canoe, and that one of the crew tried to rape an Indian woman, who vigorously fought back with her fingernails and forced him to let her go. On the basis of this evidence, we may wonder which of the two groups was the more civilized. In any event, this first encounter, although it was a complete disaster, gives us some sense of contemporary ways of thinking.

Early in the afternoon of November 14, sighting through the humid air of Saint Croix other islands to the north barely peeking over the horizon, the fleet set sail, leaving behind bitter memories, one may imagine. Columbus named this archipelago in honor of Saint Ursula and her eleven thousand virgins martyred by the Huns in the fifth century. The earliest narrative of the discovery of these islands comes from Alonso de Santa Cruz, a geographer on the expedition. The island of Virgin Gorda, first called Saint Ursula, was the first stop; the fleet then sailed through a broad stretch of sea, past Peter, Norman, and Saint John, arriving at Saint Thomas on November 17, which they named Santa Ana in honor of the Virgin Mary's mother. Bad weather prevented them from landing, but they provided good descriptions of the islands. There were no indications about the local population, but it was later learned that they were Arawak. On November 18, the entire fleet headed for Puerto Rico.

On the nineteenth, three of the twenty-five natives taken captive on Guadeloupe and Saint Croix jumped overboard, swimming either to Vieques or to Puerto Rico, telling us something of their living conditions on board and their relations with the crew.

Between the first encounter in 1493 and 1509, nine other Spanish boats dropped anchor in the waters of Saint Croix.

From December 8 to December 12, 1595, in the course of his last voyage, Sir Francis Drake stayed in the islands. Sailing from England with the mad idea of seizing Puerto Rico for his country, he dropped anchor from his ship *Defiance* first off Virgin Gorda, then crossed the strait that now bears his name between Tortola and Saint John. He then sailed for Puerto Rico, which was fiercely and successfully defended by the Spanish. The queen's pirate died of dysentery in Panama on January 27, 1596, before he could return home.

Nearly a century later, on July 31, 1683, a pirate ship was destroyed by fire in the bay of Saint Thomas. In April 1682, this thirty-two-gun ship, named *La Trompeuse*, commanded by a Captain Paine and leased for five hundred francs a month to the king of France, had dropped anchor before Saint Thomas. It was coming from Cayenne with a cargo of sugar. Paine was a Protestant and was fleeing the persecution of the Huguenots. He wanted to settle on the island and sell the ship and the cargo, which, he said, belonged to him, and he did. The French monarchy, through its ambassador in Jamaica, claimed on the contrary that ship and cargo were its property.

Two Saint Thomas merchants purchased the boat,

which was shortly thereafter captured by a French pirate named Jean Hamly. After sailing to Africa and around the islands and pillaging many galleons, the ship returned to its point of departure. At three in the afternoon on July 30, 1683, *La Trompeuse* sailed into the port of Saint Thomas with a cargo of gold, silver, and slaves. It was given protection. The next day, an English vessel, the *Francis*, commanded by Captain Carlyle on a mission of pursuit against all pirate ships, came under fire from *La Trompeuse*. It fired back and set the ship on fire as the crew escaped to land. A huge explosion made the port shake.

The governor of the island, then a Danish possession, sent a letter of protest to the captain of the *Francis*. A serious diplomatic incident ensued. Stapleton, governor of Nevis and all the English possessions in the West Indies, sent a letter accusing the Danish governor, named Esmit, of sheltering pirates, justified the destruction of the ship, and threatened to send his entire fleet. The French governor of Martinique, the Chevalier de Saint-Laurent, also got involved, sending a protest to Stapleton on behalf of the king of France in which he recognized that the vessel had become a pirate ship, but that it was originally French and should have been confiscated rather than burned. The king of Denmark, Christian V, who had signed an agreement with the English barring assistance to pirates, was worried by this imbroglio and dismissed Esmit a few months later in 1684. He was imprisoned and left the island for Denmark in July 1686. His wife Charity pleaded for him to the head of the West Indies Company in Copenhagen through the gift of a slave and secured his return to the island, but he was never again governor. When he returned to Denmark in 1689, he offered his services to the Swedish ambassador in Copenhagen to seize Saint Thomas by stealth, but nothing happened. This was not the first time that Esmit had protected pirates. Perhaps he had begun to think like them or thought he was right to collect their wealth. In any event, his lack of loyalty to his country was unacceptable. He died, forgotten, in a Baltic port.

Fifty years later, in 1733, one of the fiercest slave revolts ever seen in the islands was brewing on Saint John, which is separated from Saint Thomas by a channel barely two miles wide. This uprising, which was brutally suppressed, seriously damaged relations between blacks and whites in the region for the next hundred years.

Eight years before, after a severe drought and two hurricanes that partially destroyed the sugar and cotton plantations on the island, the owners had allowed some of their slaves to starve to death. The slaves who survived then began systematic pillaging. The harsh repression that immediately followed led to the execution of almost all the remaining slaves. In 1730, when climatic conditions had returned to normal, there were 109 plantations on Saint John, with serious manpower shortages. The Danish governor, Gradelin, authorized several slave ships from Africa to discharge their cargo on Saint Thomas. In the interim, to avoid a new revolt, the planters (like and eighteenth-century Taliban) had stiffened the rules governing slave conduct. All sorts of punishments were established, ranging from the loss of a leg, an eye, or an arm to death by hanging. Reasons for punishment might include the mere hint of raising a hand against the master or the failure of a black man to give way before a white man.

At the outbreak of the uprising in November 1733, after drought and hurricanes had again devastated the plantations, there were 1,068 slaves on Saint John. One soldier and one magistrate were killed on the first day of the rebellion. Armed with sticks, machetes, and stolen pistols, the slaves then sacked the entire island, killing some of the planters and forcing the others to flee to Saint Thomas. The first punitive expedition headed by the Danish landowners from Saint Thomas was unsuccessful. Some soldiers from an English squadron passing through the area, sent as reinforcements, fared little better. Another squadron also failed. By the end of three months, Saint John had been put to fire and the sword, many soldiers had died, and the planters on the island feared for their lives. The governor sent an emissary to Martinique to request reinforcements from the French garrison, promising in return four-fifths of the surviving slaves.

The French soldiers, led by an officer named Longueville, put an end to the rebellion in three months. The forty surviving slaves, whose lives they had promised to spare if they surrendered, were executed. Fifty plantations had been burned down. The French refused the 5,000-guilder reward, but agreed to participate in the three days of celebration especially organized for them. Ten years went by before it was possible once again to export sugar and cotton from Saint John. In fact, these events and the 1848 decision to abolish slavery brought about the disappearance of the plantations.

As though the extermination of the Arawak were not enough, this bloodshed had once again spread misfortune throughout this part of the world. I wished for the miraculous ability to go back to the past and give these unfortunates all the time in the world to tell us about what remained of their dreams.

Comfortably settled toward the stern, with a good cup of tea with just a touch of milk in front of me, I contemplated the inner sea surrounded by all those islands that had been the stage for so many tragedies. To the left was a string of islands with verdant hills just as Columbus had left them to port when he passed by. To the right, the majestic island of Tortola, long and high, still presented its green covering, fringed by yellow beaches girdling it, although the slopes are now sprinkled with many small buildings. Far in the distance in front of me, a narrow channel barely visible on the blue horizon reminded me of the passage of Drake's sailing vessel, the *Defiance*. To the left of the passage one could make out the tops of the mountains on Saint John, and on one side could be seen a tall brick chimney of a forgotten plantation. At the farthest spot of this landscape peopled by memories, beyond the strait, could be glimpsed the pale mauve heights of Saint Thomas. If you squinted, you could even make out delicate puffs of white smoke causing balloons of mist to rise slowly in the blue sky. The smoke was not coming from the *Trompeuse*, but from some growers burning their dead trees to clear their fields. Seen from a distance, nothing had changed very much, and the bright blue color of the water was certainly the same as it had been long ago.

And yet not really. Today there were fifty carefree yachtsmen tacking across the glittering blue water who hadn't the slightest idea that other sailors, long before, had traveled thousands of miles to discover this place one fine day in 1493, over five hundred years ago. Was the saying attributed to Socrates true, that on this earth there are the living, the dead, and those who go to sea? Where are all those departed souls today?

As though to compensate for the darker moments, this internal sea that had witnessed the tormented past of men and women, many of whom had faded from memory, had also surely been the scene of charming love stories. No doubt one or two of Longueville's French soldiers who joined in the celebrations had been taken with the charms of the blond-haired Danish women. And don't forget the story of Rose and Eddy, surely only one of many.

It was time to raise the anchor. We needed to refill our diving and fuel tanks and send our recordings to Florida. The marina of Yacht Harbour, near the Baths, would suit our purposes. Two days later, after watching a cricket match on a nearby pitch, I was to take a ferry to Tortola. I wanted to be sure that the FedEx for Florida would go off as soon as possible and that the scheduled arrival of the plane from Saint Martin had not been changed.

Road Town, the capital, was swarming with cars.

Since our departure, I had almost forgotten traffic jams, dust, and the noise of automobile horns. The market, not far from the jetty, was covered with stalls protected from sun and rain by multicolored awnings. My attention was drawn not to the bright colors of the exotic vegetables with unusual shapes, nor the spicy aromas emitted by the island plants, but rather to the languorous eyes and tender smiles of the buxom black women who proclaimed the quality and freshness of their produce. They had left their villages perched on the hills overlooking the emerald bays early in the morning to bring to market the products they had carefully grown on their little plots of land. Farther on, the imposing buildings of the banks and government offices had left little room for fashion and fabric shops, whose owners, descendants of East Indians brought here as laborers in the last century, had remained as tradesmen. Throughout the islands, it is not uncommon to see on shop signs such names as Dadlani or Mathaani.

A postal employee told me where to find the FedEx office.

Our next mooring was in the center of the sandy bay, well protected from the winds and bordered by coconut trees, located on the north coast of Peter Island. From this anchorage at night, just after sunset, you could see the island of Tortola illuminated by a thousand tiny points of golden light, reminding us of the recent building boom there. Unfortunately, the sea grass beds just beneath our hulls were deserted. We would not be able to pursue our listening here and perhaps nowhere else in the islands of this archipelago. But who knows? In the meantime, we would relish the idyllic setting, for it was not certain we would find any more beautiful in the course of our expedition.

The many pleasure boats anchored before Norman Island blocked the approach to the treasure caves. The legend is that the caves were for a time used as hiding places for pirate booty. A brief dive among the multitude of fish that were found there, who fed on bread and other food thrown to them from boats by vacationing tourists, led to a troubling discovery. Many fish, such as the yellow-tailed sardes, for example, had brown tumors on their bodies, probably indications of cancer. Later, when we stopped at Saint Thomas, we told researches at the marine university about the anomaly; already aware of it, they were taking samples and conducting analyses.

On the other side of the channel that now divided us on the west from Saint John lay one of the most beautiful peninsulas that we had seen. The many wooded hills that towered over it were devoid of any buildings. The shoreline was very jagged and offered wonderful shelter from storms.

Dropping anchor in one of these bays, we had the impression that we were alone on the far side of the world. Perhaps that was the explanation for the name given to this part of the island, East End. We chose this deserted spot not only for the countless mangroves that bordered the narrow beach encircling it, but also because of its name: Prince Bay perfectly suited our *Prince* and us as well. The multitude of blue herons nesting in the trees that lazily flew off and settled a little farther away when we arrived told us that there were alevins beneath the murky waters in which the high roots of the mangroves were growing. Indeed, if I had been a fish, it was here in this sheltered spot that I would have come for my mating dances. Our hydrophones proved us right. As soon as we dived, just before sunset, beneath the quiet surface into this extraordinary love nest, the concert began. All the muffled staccato sounds indicated the presence beneath us of many gray porgies. The duration of the repetition of the sounds and the consistency of their strength led us to conclude that we were witnessing, through the hydrophones, one of their nights of love. The full moon dusted the surface of that magical cove with thousands of golden gleams and seemed by its cosmic presence to offer the gray porgies its blessing for the continuation of their species. In fact, we could see the moon smiling. The same thing happened several nights in a row, providing a good deal of material for our minidisks and our notebooks.

Professor Gilmore's team at Cape Canaveral must have been satisfied with our work. During one of my visits a few months later, after he had read my notes, he asked me countless questions about that extraordinary night. The darkness of night, by favoring sonic signals, proved once again that there were exchanges of information between fish of this species. Were the sounds only to indicate their presence in the darkness to their neighbors, or did the sounds from the males indicate their availability to the females, or the sounds emitted by the females reply to their advances, or all of that at once? Only NASA's computers, by analyzing the frequencies of the sounds and comparing them to others, could provide the beginnings of an answer.

The next morning, with a following wind filling our spinnaker, we crossed Drake's passage and headed for the island of Jost Van Dyke, once owned by the Dutch pirate whose name it bears. This little paradise still gave off the odor of gunpowder from the ships of pirates and smugglers that followed one another here. Each of its bays, lined with fine sand and shaded by a row of tall coconut trees, had no doubt been the lair of brave and reckless men wearing, if the engravings of the period are accurate, wide blue canvas jackets, coarse cotton trousers with gold buttons, broad-brimmed black hats, and red sashes around their waists in which were stuck two flintlock pistols with the skull and crossbones, and a sword with a protected hilt. One of their vessels, a little sloop with a long bowsprit and the rigging of fishing boat — two masts of Latvian wood rigged with triple sails — with hemp stays and ropes, creaked impatiently as it rotated around its fluked grapnel.

But you shouldn't place too much stock in my tale, because the grapnel was used only to catch an enemy ship in the rigging and pull it alongside. So the boat really rotated around its double-pronged anchor. Sitting athwart the yardarm from which hung one of the square sails, a sailor with his shirt open showing large scars constantly watched the horizon, with his one remaining eye . . . I think you're following me now.

After tying up at one of the buoys set off the coast, we approached the white sand beach in the dinghy. To our right, up against the hill and hidden beneath some frangipani trees, was a hut made of driftwood surrounded by broad awnings covered in reeds, held up by roughly carved posts. A few tables set beneath the awnings were an invitation to drink a cold beer. On the roof was a large black and yellow sign that read "Foxy's Bar." To the left, sheltered from the sun by a row of royal coconut trees, was a row of rustic houses with weather-worn gray wooden walls and palm roofs darkened by the sun. These flimsy little houses seemed to have been set there by pure happy chance on a delicate carpet of greenery. In the center, beneath the oppressive heat of this Sunday afternoon, the red corrugated tin roof of a little church whose fragile spire barely rose above the tops of the trees seemed to be both a rallying point for the parishioners and a seamark indicating from the open ocean the entrance to the narrow channel leading into that magnificent bay.

The tinny sound of an old radio broadcasting religious music drew me to one of these rustic cabins. I made my presence known with a loud hello. Through the open door I could see an old woman sewing. She was listening to what sounded like a violin, with the bow drawing from the strings the moaning sounds of an old West Indies tune with a slow rhythm. Not daring to disturb this peaceful seamstress in her work, and daydreaming on the threshold, I tried to identify the melody.

Then, looking up at me, she put a finger to her lips to ask me to remain silent and keep listening. Whether it was the damp warmth of the comforting shade of the coconut trees, the peace emanating from the place, or the slow rhythm of the chorus that gently took the place of the roughly tuned violin, something made me imagine for a fleeting moment that I had always been a part of that natural setting. It was a kind of trance in which it is impossible to tell where one is. I had the impression of belonging simultaneously to the trees, the beach, the entire landscape, almost a feeling of immortality. I was the sky, the clouds, the sea, the hills, and the undergrowth all at the same time. When the music finally stopped, I returned to earth. The old woman smiled at me. I had no idea whether she sensed my state of feeling.

"Hello, my name is Gilles."

"And mine is Lucy," she answered.

"That was very pretty, what you were listening to."

"It's an old radio, you know. It keeps me company, especially on Sunday when it broadcasts services from the other islands."

"Do you know that tune?"

"Yes, it's a group from Saint Croix that sings church music, but I don't know its name. It may be "Shine the Light," but I'm not sure. Anyway, they're on every Sunday."

Our conversation lasted a few more minutes and then she went back to work. I thought that she, too, had shared those unreal moments. Let the light shine and to everyone his own happiness.

Two months later, when I passed through Saint Croix, I spent more than a day searching for this group. Finally, after unsuccessfully looking through all the record shops on the island, by chance I heard the same tune when I passed in front of a chapel in the port. One hour later I was in possession of the CD. Listening to it on board months later, I was reminded of the peaceful West Indian woman sewing on a Sunday afternoon, listening to the radio under her palm roof. This pirate island had over the years replaced the rough songs of passing sailors with West Indian gospel music. Lucy probably understood everything about life.

There was a spot on the map far from all traffic that seemed to be of interest for us: a little island north of Jost Van Dyke called Green Cay, which, protected from the ocean by a coral reef nestled between two other islands, might provide the calm necessary to continue our work.

But this failed to allow for the bad weather that began the following day. The rain that had spared us since Barbuda surprised us as we were mooring before the island. The spot was magical despite the strong gusts of wind that complicated the maneuver. Making a false step and sliding on the wet deck, I displaced one of my vertebrae or one of my disks. Totally immobilized, and forced to find a safer mooring, I decided to drop anchor in the very well protected bay of Cane Garden on the west side of Tortola. Ten days without moving. Not knowing exactly what was wrong with me, I felt jealous of all the people in the village at the end of the bay who could go about their daily lives.

With the help of anti-inflammatories that my doctor brother had gotten for me before sailing, I gradually came to the end of my suffering. For the first few days I thought that the expedition was over. I felt unable to face again the movement of the boat tossed by the waves or to jump into the dinghy to gather our recordings. My morale had sunk to the bottom, and I felt miserable. Helped by my crew, who served me meals in bed, my only distraction was to prop myself with difficulty on my elbow to look through the porthole at the incredible spectacle of the pelicans. Like stunt pilots flying in formation, they took off in lines and plunged in showers onto the surface to capture, with the help of the thick pouches under their beaks, as many fish as possible, who were no doubt terrified by this flying shadow and tried to escape.

I had all the time in the world to admire their technique. Scouting was essential. Flying as high as possible, these feathered Concordes, after detecting their prey, slightly turned their beaks toward the surface and began a breathtaking descent that seemed to risk breaking their necks. As they came down, their wings were partially folded back, so that at the moment they hit the surface, the wings were fully aligned with their bodies. Not a trace of foam indicated the location of their dive. They could have won all gold medals at the next Olympic Games. Their tactics resembled those of the gannets that we had observed on Dog Island. The speed of penetration seemed to be the same. But the gannets stayed longer underwater and immediately flew off on returning to the surface carrying their prey gripped in their beaks, whereas when the pelicans came out of the water, they stayed on the surface and swallowed their fish with a large neck movement, and then took to the air again.

Lying in bed watching the agility and freedom of movement of those birds, I wondered if I would ever be able to walk again. After a week of anxiety, there was finally some progress. With the help of improvised crutches, I was able to return to the rear deck for meals. The effort tired me out but it gave me hope. Fortunately, this mooring was therapeutically calm. Another few days for recovery and we would return to sea. We had to get to the other side of the island to get closer to the airport.

One day was enough, sailing into the wind, to reach the northern coast of Tortola. At every sudden movement of the boat, I silently worried about my back. Would it be able to handle the situation? The channel separating Tortola from Guana was as majestic as Drake's. Between that island and Camanoe, I sighted from on board many deserted spots where we could have continued our research. But I now had to be patient and not let frustration trouble my mind.

From Trellis Bay, where we were now comfortably anchored, we could see the control tower of the little airport.

All around the bay, on the edge of the beach, were a large number of shops belonging to fishermen from the island of Dominica. Half sculptors in wood, half builders of dugout canoes, they were finishing the construction of a large canoe that was going to sail from Tortola to the mouth of the Orinoco on the coast of South America. This expedition, which would be fascinating to follow, wanted to show, in the opposite direction, that the Arawak had used this means of navigation to settle the islands of the West Indies archipelago two thousand years ago.

The Arawak, armed with the knowledge they had gained from cultural exchanges with other tribes made possible by the ease of communication on the great river, had fled from the persecution of the thieves of souls and ideas that were the Caribs to settle in a better world. To bring this hope to fruition, they had sailed with women and children and a few plants and animals on this great adventure. With the strength of their arms and their paddles, probably helped by the prevailing southeasterly winds, they had moved north from camp to camp. Having arrived here near a water source, they could flourish, creating one of the most advanced civilizations of the era, planting the manioc of their native land and developing medicines that allowed them to live in peace until a certain Christopher Columbus arrived in their waters. You know the unhappy results.

To return to the fishermen from Dominica, they had reached the final phase of their construction. They were sculpting on the prow one of the most beautiful bird figures that I had ever seen. It was the head of a toucan, the bird with brilliant plumage and an enormous beak native to the banks of the Orinoco and all of Amazonia.

The Winnair plane from Saint Martin was for once on time.

Our guests, surprised by the heat and humidity, quickly stowed their things on board so they could take advantage of the waters in the lagoon in which the *Prince* was moored. I was very pleased to see my brother and sister-in-law after all the months of adventure and research. After a good welcoming dinner in the restaurant decorated like a pirate den built on a rocky promontory called Bellamy in the center of the bay, the doctor's diagnosis was short and sweet: rest, rest, rest.

For the following days, until they left for Puerto Rico two weeks later, our guests were placed on the crew list and followed the tourist itinerary that we had organized for them. We had planned, because we had picked them out during the month we had spent in these waters, to visit, from mooring to mooring, the most beautiful bays and beaches of the islands. We had a plethora of choices. The Baths, Fallen Jerusalem, Peter Island, Norman — all the islands they would discover as Columbus once had.

In the bay of Jost Van Dyke, my seamstress was no longer there, but Foxy's beer was still just as cold. The turquoise waters of Green Cay in which we had been unable to work during our first stay because of my mishap gave my brother the opportunity to shoot a lot of film. The sun that day changed my perspective on the unfortunate memory. The coconut grove set in the middle of this minuscule island, surrounded by a beach of bright yellow sand lapped by clear emerald waters, was enough to make the shutters of his cameras open with pleasure. The bay of Cane Garden on Tortola, which two weeks earlier had been the scene of my forced rest, was also an excellent memory for them. By this time, the state of my back had distinctly improved, and we set off to climb the highest peak on the island, towering 540 meters over the entire archipelago. Before reaching the goal, we had to go through a forest of cedar, gum, mahogany, and iron-wood trees. Because the last had been used to excess for ship-building, its use is now forbidden and it is one of the species protected by the government of Tortola. The narrow steep path was bordered by tree ferns, known as elephant ears, with very broad leaves often blocking the way. When we reached the platform on which a rough observation post had been set up, we were treated to a magnificent panorama. In the distance to the east we could glimpse the heights of Virgin Gorda, barely masked by the fine heat haze. To the south, bordering the indigo of the inner sea, lay a green necklace of dozens of islands with enchanted harbors. The island of Saint Thomas, off to the west, was also completely visible. Its mountains, tinged with mauve, were clearly set off from the transparently pale blue horizon. This "watercolor" landscape, washed by a cool breeze left behind by a front from the north, almost made us forget the stifling heat of the bay that we had left below us. But our descent on the back of a pickup truck that had given us a lift soon reminded us that, down below, heat and humidity had not miraculously disappeared.

Although the *Prince de Vendée* flew an American flag because of my dual citizenship, we learned from the harbor police that we could not enter American territory with passengers carrying foreign passports, even if they were on the crew list. No problem — we would drop our guests at the Soper's Hole ferry, which goes back and forth three times a day between Tortola and Saint John and is authorized by the American immigration authorities, where they could get all the necessary forms for entry into the country, notably the form to obtain a visa.

After crossing into American territorial waters on two different boats, we rejoined forces in the little port of Cruz Bay, capital of the island of Saint John. The *Prince*, for his part, had no problems. He was returning home.

Having become in large part a national park in 1954, Saint John also has sumptuous harbors for boats. The thick vegetation that covers its hills long ago erased the traces of the old cotton plantations. In the streets of the capital, the only visible rebellion was that of the drivers who fought to fill their buses to take tourists to what they said was the most beautiful beach in the world: Trunk Bay. This was probably true. I chose a mooring facing Cinnamon Bay. It had the advantage of being not far from the capital, making it easy for our guests to shop, but also I could meet the owner of the hotel on Caneel Bay who had welcomed me three years earlier. This was an old plantation that had been the last refuge for the planters during the 1733 revolt.

Two days later, a wind strong enough to fill our jib took us to Saint Thomas. From offshore, you could still detect the damage caused by Hurricane Lenny in 1999. Many formerly wooded mountainsides were still without vegetation. With hurricanes Luis and Hugo, this island has had more than its share of misfortune in recent years.

That day, the huge protected harbor of Charlotte Amalie was filled with the most beautiful cruise ships in the world. I was pleased to see the *Queen Elizabeth* again, on which I had crossed the Atlantic a year earlier. The *Norway*, formerly the *France*, was moored offshore. There were still others at the dock. All these boats discharged on average seven thousand passengers a day, who rushed into the duty-free luxury shops to buy perfume, liquor, jewelry, and souvenirs, which they would have a great deal of trouble fitting into their suitcases when they took the return flight from Miami. Squeezed between these white-painted walls of steel, our boat seemed quite small.

During this stay, I met researchers from the Marine Science department of the University of the Virgin Islands, to tell them about our work and exchange information about the behavior of groupers. They also complained about the disappearance of their spawning areas and were a little jealous of our equipment. One of them, Mike, invited me to participate in his research the following year.

In Fort Christian, which overlooks the town and was familiar to me because of the many visits I had made in the past, I discovered by chance an old engraving of the *Trompeuse*. A group of divers whom I asked if they had tried to locate it told me that an expedition intended to recover its remains was under preparation in the strictest secrecy. Passage through the island of so many pirates had left a few traces in everyone's way of thinking, but that hardly mattered, because everything here recalled their adventurous lives.

My sponsors, reassured to see that I was still alive, asked me countless questions. I described for them the research program for the coming months. It would be concerned with volcanoes and tectonic plates. I gave a lecture for their clients on the island and participated in a television broadcast. They informed me that they were also in contact with NASA, which had given them their compliments on our work, and they confirmed their financial support for the year.

Meanwhile, our guests were filling their suitcases with

souvenirs. Taking advantage of the abundance provided by the supermarkets, and particularly of the wonderful prices compared to those on neighboring islands, we took on board all the provisions we could.

I left the island feeling very nostalgic. Once in the open ocean, seeing on the heights the old French quarter of Malfolie and the summits of the mountains protecting the harbor of Charlotte Amalie recede behind me, I thought of all the friendly pirates that I had met over the course of more than twenty years and that I had been able to see again with great pleasure. But other Virgins were awaiting us beyond the horizon, and they were Spanish.

8

Of Rain, Bullets, and Bombs

A burst of wind — An unknown paradise — Populated depths —
A man from Vendée passing through — Rain in May —
The fortress on Puerto Rico — A stop full of encounters —
The islands on the east coast — The island of Vieques and bombs
— Lights in the night — The Coast Guard —
An easy crossing to Saint Croix.

That morning, in fair weather, we were thirty miles from the first of the Spanish Virgin Islands. A large, docile flock of cumulus clouds scudded before the trade wind. We estimated that the moderate east wind, producing a gentle swell, would enable us to cover the distance in four hours. The spinnaker had been removed from its pouch so that it could be unfurled from the forward points of the boat to the top of the mainmast. Its 140 square meters, with the colors of the Bacardi Foundation, propelled us at a speed of seven knots. The prows of our two hulls gallantly cut through the surface of the ocean, leaving behind two identical straight wakes. Some of the crew were sunning themselves on the bow while others were carefully following our course on the electronic plotter set into the card table in the cabin. The aroma of fresh coffee drifted through the air, and Rachmaninoff's Third Piano Concerto charmed us with its romantic flights of melody. It was one of those rare moments when sailing is pure pleasure. The heights of Saint Thomas had just disappeared over the horizon, giving us the impression on this empty surface that the *Prince de Vendée* was the first boat in the world to sail on this docile ocean.

The calm confidence we momentarily reposed in the weather failed to account for its capricious character. It was already early May, a time of year when weather fronts reverse. Low-pressure cold fronts coming from the west early in the year were now coming up against a solid front of trade winds from the opposite direction and were gradually losing ground. This high-altitude battle could have only one winner. This was why, for the last fifteen minutes, a band of rather worrying black clouds was approaching us from behind, from the east. It would probably catch up with us very soon.

It was time to take action by pulling on the halyard to take the air out of the spinnaker so that we could furl it and then unfurl the jib halfway so that we could await the whims of the weather without panicking.

A heavy curtain of warm rain suddenly enveloped us. The jib, now completely furled, left us without sail, and the motors in the rear housing carried us through the sheets of water. We thought that this dark curtain, after pouring its contents down on us, would soon part to reveal a scene whose backdrop would once again be painted blue, strewn with pretty white clouds with irregular outlines shaped by the breeze.

But the dark play being enacted on this grandiose stage, which we hoped would have only one act, had not yet ended. Visibility was zero. Blinded by the rain, we could no longer even see our two prows. The steely surface of the sea, tinted charcoal gray, furiously pelted by large raindrops that rebounded in thin silvery candles as though to decorate an enormous cake, reflected the dark shadow of an unreal sky. How long would this last? We were probably not far from the irregular and now invisible coast of the first island in this archipelago, with which we had little familiarity. To avoid a last-minute engine failure or being swept away by currents, we headed back out to sea where we could be free of possible dangerous encounters. Our European crew, for whom this was their first ordeal at sea, cast a worried eye on the goings and comings of their captain between the navigation instruments and the tiller, but nonetheless approved his wise decision. Suddenly, while all the members of the crew were shivering even though they were protected by yellow slickers, with nervous smiles on their streaming faces, the curtain parted to reveal a clear horizon.

The tender green hills of Culebrita, the first of the Spanish Virgin Islands, appeared in the distance, at the base of the comforting backdrop. The Rachmaninoff concerto had just come to its quiet end.

This little island is surrounded by many obstacles. We had made the right decision to bide our time on the open sea.

The narrow "Tiempo" channel is hard to enter. It

snakes between Genequi and Cayo cays to the north and Botella Cay to the south. An entry in bad weather into the tiny protected bay would have been very risky and probably catastrophic.

Now that we were safely anchored facing a smooth band of still damp sand bordered by a dense row of mancenilliers, the memory of the heavy curtain of rain faded in an instant. Made hungry by all the excitement, we quickly prepared a meal.

This immaculate island is a strange spot. A veritable paradise devoid of sailors lay before us. At the top of the only hill, covered with spiny shrubs and cacti, stood the tower of an abandoned lighthouse. To our left, a pool of seawater was set amid a maze of rocks, and at every wave it became a swirling natural Jacuzzi. The contrast with the other Virgin Islands was striking. Here it was difficult to imagine the presence of human beings in the past; it seemed to be virgin land, a setting for a film that would stimulate the winter dreams of searchers for treasure islands. A long tracking shot opening the film, taken from the top of the hill and gradually showing all the jagged outlines of the island, would soon take your mind far from daily concerns.

As we were already from the very first image, you, too, would be cast back to your childhood dreams. Was this the place where the three Indians, two men and a woman, who had jumped overboard to escape from Columbus five hundred years ago, had managed, swimming through the currents with vigor and courage, to come to land? What a great idea for a film. These Robinson Crusoes would have adapted to the setting, as they already had on other islands, building a simple shelter and subsisting on fish. Against their will, they would have come into conflict with the tribe on a neighboring island. One of them, attracted by a pretty Indian woman, would stay there, while the other couple would return to their native island on a canoe they had built, to recount their adventures to the worried families they had left behind after the passage of the Spaniards. But since such a movie doesn't have enough suspense, I'll probably never see it.

The general map of the islands indicates that the neighboring island, Culebra, is surrounded by many deep bays protected by coral reefs. The channel to its south, which separates it from Culebrita, is deep enough to sail in. It is bordered on the east by a long coral reef that continues to the south in an extension of the island, and on the west by high cliffs marking the edge of the plateau on the heights of Culebra.

We dropped anchor first in the middle of Almodovar Bay. This huge natural port, which you enter through a quiet, short, and narrow channel, was of even greater interest than our "studio" on Virgin Gorda. To port was a long row of mangroves on which delicate egrets were nesting, and to starboard a fine line of spray, barely rippling over the calm surface of the large lagoon, indicated the presence of a coral reef. We were protected from all the winds. We were alone.

With the consent of the ship's doctor, the research work could resume. Two days and two nights of listening confirmed our earlier discoveries. Our new acquaintances exchanged information through sound twenty-four hours a day. Although not continuous, their dialogues did not stop at sunrise. As I did at our next stop, I recorded here hours of sounds that demonstrated this fact. The same sonorities, with different intensities, always meant either territorial defense, recognition of the presence of danger, or a call to females followed by their replies. The difference for me after three months of experience was that I could recognize certain species by the sounds they produced. I remembered my bird-watching trip to the cliffs in the south of the Monterey Peninsula. I was not far from achieving what I had dreamed of at the time.

After working here, I was inwardly grateful for all those who had in 1975 made these islands an enormous nature preserve, a place where hunting and fishing were prohibited and nature had the opportunity to resume its ancient rhythms.

Little is known about the history of these islands. With a little imagination, however, and recognizing their virgin state, one might think that they have been sheltered from human folly from the very beginning. Perhaps some pirates hid their booty here.

The *Prince de Vendée*, no doubt out of regional patriotism, whispered that in these waters a celebrated native of the Vendée born in Sables-d'Olonne in 1630, on board his vessel *La Poudrière*, a 100-ton ship with a crew of forty, is said to have sunk a Spanish galleon with his twenty-kilo cannon balls and recovered its sixteen cannons and 40,000 gold ingots.

François Nau did indeed sail in these waters after embarking in the port of La Rochelle some time in the 1650s. A buccaneer who was made into an honorable privateer by a decree signed by the governor of Tortuga, he pursued anything that had a sail and floated, seized his prizes, and took them to Santo Domingo. He is said to have been killed and eaten by Indians on the coast of Nicaragua in 1670. The collective memory of those Indians must have surfaced from the past, with reminders of the harsh treatment inflicted on them by the army of Cortés more than a hundred years before. But the *Prince*, in Barbuda, Saint Barts, or Anegada, sees his compatriots everywhere.

Making a few dives at the coral reefs of this archipelago, I was pleased to see that they were all in perfect health. Protected from the strong waves of the open sea, they presented a huge variety of colors and life forms that I would have loved to see elsewhere throughout this journey.

The little port of Dewey, nestled in a deep bay, seemed to be the favorite holiday spot for natives of the island, who earn money in Puerto Rico and then return here for vacation. The charming restaurant, with the reassuring name Mamacita, overlooking the Luis Peña Canal, a narrow shortcut to reach the open sea from the lagoon, was chosen as our last stopping place before sailing for Puerto Rico.

Moving off from Tamar Point, which dominates the west coast of Culebra, we set a heading of 275 degrees toward Puerto Rico. The Luquillo mountain chain was clearly visible from our shipboard vantage point. We were only eighteen miles from the Del Rey marina where we were to spend a few days. To starboard we were sailing past a long coral reef topped by little desert islands forming the northern border of the basin of the Vieques Islands.

After passing the Virgin Islands during his second voyage, Columbus saw Puerto Rico for the first time. This large landmass set on the horizon, which he saw from a distance, as we did, must have made him think that he had finally discovered the Indies.

The huge sanitized marina with long fibrocement docks had a comfortable spot for the *Prince*. We had to use electric carts to get to the port office.

The rain that began the next day remained with us for the rest of our stay. On the road to the capital, I counted no fewer than 120 self-service restaurants. American fast food had probably ruined the typical little island roadside restaurants.

Visiting El Morro, the citadel constructed on several levels of ramparts, on which are set cannon batteries to protect the old city of San Juan from attacks from the sea, I realized that two hundred years ago it required huge firepower to capture it. In 1595, with all the gold of Mexico, the Spaniards built a battery comparable to the one in Málaga, and it is easy to understand how hard it was for Sir Francis Drake to capture it. Its outer walls, six meters thick and reaching as high as forty-three meters above sea level, made the fortifications impregnable.

On the ramparts with overhanging sentry boxes, set in the embrasures of the crenelations, can still be seen the powerful cannons with their imposing piles of black cannonballs. From there, in good weather, you can see the outlines of the peaks of Saint Thomas.

After being driven out of Cuba by Castro, our sponsors the Bacardis had set up their rum-making operations in Puerto Rico. A press conference was organized while we were there. In view of the adventures awaiting us, they filled part of our hold with the comforting, much-needed beverage.

After walking beneath the green vaults of the unique

tropical forest of El Yunque, which towers 760 meters over the entire east coast of Puerto Rico, with a heavy heart we saw our guests to the plane taking them back to Europe. Their return trip to France via Saint Martin was delayed by missed connections. Twelve hours later, with their eyes still dazzled by the sun, they reached La Rochelle.

Simon, our loyal sailor, regretfully left us as planned to return to college in Maine. He sent an e-mail on his arrival thanking us for everything he had learned on board. Raphael, whom we had recruited during our stay in Marigot, arrived just in time to take his place. His first task was to put the boat in order and to verify the condition of the engines, which he did very well. Raphael, always available and competent, was an indispensable resource for the expedition.

In the meantime, I spent several hours every day in front of a computer screen in the marina office in order to coordinate by e-mail the arrival of the two researchers who were to come on board in Saint Croix. They worked in geophysics laboratories in two American universities, one in Chicago and the other in Houston, and they had been commissioned by Professor Dixon to go with us to take measurements of the movement of the Caribbean plate on the island of Las Aves. Another pair, recruited by Professor Dixon in Miami and Professor Mattioli in Washington — also geophysicists — would meet us on the island of Dominica to continue research on volcanoes with the instruments we would take on board on Saint Croix. I also learned that a geologist from the University of California at Santa Barbara, Alan Smith, would come to Dominica to join the team.

My solitary research would temporarily be suspended.

Professor Gilmore of NASA, who had promised to join us to listen to fish at Anegada, was detained at the last minute by a serious family problem. From that time on, I received at least once a week his comments on the minidisks that I regularly sent him. He directed my work every week by e-mail.

The military forces on the island finally gave us authorization to anchor at Vieques, a training area for the U.S. Air Force. The authorization was limited to certain days of the week and to particular spots where there would be no danger of being accidentally hit by one of the 500-pound bombs that the air force regularly dropped in its training exercises on the eastern end of the island.

Since the upcoming maneuvers would end in a few days, we had a little time in front of us. We left the marina in a driving rain, happy to leave behind the soulless concrete piers.

Two little islands, Palomitos and Pineros, were the setting for a few days of research on the sounds of fish. Unfortunately, although from their position on the map they seemed to offer all the necessary characteristics for this kind of work, these islands were surrounded by ocean water made murky by the alluvial deposits carried by the rivers flowing down the slopes of El Yunque. All the reefs had been affected by the phenomenon. Their system of reproduction was blocked by the impurities floating in the water, and they were in the process of swiftly dying. As a result, we were a little ahead of schedule when we rounded Arenas Point at the extreme west of the island of Vieques.

This long hilly island has many deep bays on its southern coast suitable for the reproduction of fish. Most of them have dense mangrove forests around their perimeters. Because of the danger of the often-not-very-precise bombing on the eastern part of the island, that part is normally completely deserted, except for local fishermen who venture there with the authorization of the military.

At our first stop in the bay of Puerto Real opposite the little port of Esperenza, to familiarize ourselves with the habits of the fishermen and to make their acquaintance, we were told that most of them lived on the other side of the island, in the capital, Isabel Segunda.

The minibus that took us there was stopped halfway by a large demonstration. Hundreds of men, women, and children had gathered to march in front of a barbed-wire fence which we took to be the border of the eastern bombing zone. Hung on the fence dividing the island in two were a large number of cloth and plastic banners with all sorts of anti-bombing and anti-American slogans. Also set before the fence were a significant number of white crosses memorializing accidental deaths from the bombing and two dolls on a white sheet on the ground surrounded by blue plastic flowers. From the rear platform of a truck, one of the demonstrators harangued the crowd through a bullhorn. Words like "freedom" and "independence" recurred frequently. Official representatives from the whole island were there. The mixed and angry crowd repeated the slogans, brandishing their fists. On the other side of the fence, a military truck surrounded by powerless and stoic soldiers confronting this storm of demands were waiting, with the motor running.

Our driver, Pedro, an open partisan, told us that another child had died the day before from cancer caused by the toxic chemicals the bombs disseminated into the island's atmosphere. The month before, a fisherman had also been accidentally killed by a bomb, leaving behind a widow and three children. Too much was too much. This had to stop immediately, or they would demonstrate in front of the White House with their representatives next month. Violence, which had been avoided up to now, could break out at any moment. The crowd wanted to retake possession of the whole island. Twenty deaths in five years was no longer acceptable. I learned that this was not the first of these demonstrations. There had been many others before, but although the military and political authorities understood their demands, nothing had changed.

It was not until a year after our return to France that all bombing stopped on the island and it finally recovered its calm. The pacifist inhabitants had won their battle. The families that had lost loved ones were financially compensated. But money never washes away tears.

Pedro had a brother who was a fisherman in the southern part of the island. We met him after the demonstration. Juan, who was about forty, had a boat anchored off the village of Esperenza. Tall and thin, with a lively air, he looked like an overgrown adolescent, even though the hair at his temples was going gray. Clever and well educated, he was responsible for the protection of Mosquito Bay, which had not yet been made a nature preserve. He worked part-time for the Vieques Conservation Fund, and supervised boat trips in the bay on behalf of Mrs. Baker — a biologist specializing in bioluminescence — who was at the time traveling in the United States. He was to be our guide and became our friend.

There are only five or six places left in the world where the emission of light by living organisms still exists, and we intended, under his supervision, to attempt another experiment.

Mosquito Bay enjoys twofold protection from the open sea. First, there is a thick coral reef providing access only through a narrow channel, and second, a bottleneck, a kind of gorge bordered by steep hills leading to a lagoon surrounded by a dense mangrove forest. The shallow sandy bottom, from the entrance to the shore of the bay, is carpeted with long marine plants with flat stems and rounded extremities known as turtle grass. A few buoys set in the middle of the bay provide moorings. It was here, having been granted special authorization, that we and Juan awaited nightfall. While we were there, we could not flush out our dirty water or throw anything overboard.

"Is this the first time you're going to see this?" Juan asked.

"Yes, it is."

"We have to wait for complete darkness. The little glimmers of light you'll see on the surface are not reflections of moonbeams."

"I didn't think so."

"No, the microorganisms here use a kind of whip for locomotion. With the whip, which contains chlorophyll, they make their food with the daylight that they use for energy."

"This is the photosynthesis that corals use to make their skeletons, isn't it?"

"Yes. They store up the light just as bacteria living in

symbiosis with certain fish sometimes do and then emit it later."

"Why do they do that?"

"It's an advantage that they have over other species. They use it either to locate their prey, or to attract females, or to blind an enemy. In any case, they use the light only when they are in motion and only at night."

"Might it also be a way of communication, like sound?"

"It's possible, but no one knows. We are not even certain how the energy is transmitted to the molecules called luciferins. It's a bit more complicated than an electric bulb, which transmits the energy supplied to it to the filament that then provides the light."

"Yes, I understand."

"Mrs. Baker says that enzymes that need little energy are the source of bioluminescence."

Juan knew his subject well. It was a pleasure having him on board.

"Do you think I can put my hydrophones in the water."

"Yes, it would be interesting to see what happens."

Night fell swiftly. Our hydrophones very clearly recorded the characteristic sounds of the shrimp below us, but nothing else. We didn't really expect to hear the noise of the microorganisms, although we were on the lookout for other unfamiliar sounds. Juan, who was unfamiliar with the technique, kept the earphones on for some time. Once the experiment was over, he asked us to go into the water. Raphael, the bravest, dove into the dark water first.

The view from on board was miraculous. He created behind himself a long wake of phosphorescent light as a fairy might have done scattering sparks with her magic wand. This trail of golden powder followed him for a while and then faded into the darkness of the water. When he came back to the surface, Juan asked him to float on his back while moving his arms and legs. He looked like an angel fallen from heaven wearing a cape with millions of tiny glittering stars.

Everyone took turns for a brief dive into that phosphorescent world.

Juan continued, "It's because of the vitamin B_{12} synthesized by the bacteria living in the mangrove roots that the microorganisms are here. They feed on the vitamin."

"The ecological balance is very fragile here, then."

"Yes, that's why there are practically no other bays like this in the world. The tide can't be allowed to wash the vitamins away."

"That must be why the bay still exists, since it's so strongly protected from the sea and surrounded by thousands of mangroves."

"Yes, precisely. Do you know how many microorganisms there are in each liter of water here?"

"No idea."

"A hundred and fifty thousand!"

We can only hope that the bay is very soon declared a nature preserve.

On our way back, after ascertaining that we had authorization from the military, Juan advised us to moor for the following days farther east in a bay that he showed us on the map. He told us that this bay, protected on the west and south by two dangerous cays, was very deep. "To the right of the broad entry channel you will see a line of reefs that you can go around to starboard and drop anchor at the foot of the cliffs. The bottom is clear and in good condition. You only have to be careful of the currents, which are rather powerful near the island, but once there you'll be well sheltered from the east wind."

That's where he fished for tarpon and stopped before heading out to a spot in the ocean that he called the "grouper hole." He gave its position as 17°07' north and 64°11' west. This must be a spawning ground, which human activity had driven from the bay to a deeper spot farther offshore where the groupers could mate and reproduce in peace and quiet.

The spot suited us perfectly, even more so because it was on our way back to Saint Croix, where we were headed next. We could make our recordings in peace and perhaps capture the sound of tarpons. They generally swim in schools of twenty to thirty. They move fast and prefer to seize their prey during the day amid the currents. But in the evening, they always pick a quiet spot for the night. Weather permitting, we would then set out our hydrophones farther offshore among the mating groupers.

Juan showed us around his village and then came on board for dinner. We took on some provisions at the same time. Early the next day we set sail. I wanted to leave early enough so that we would have good visibility when we arrived.

The north wind that propelled us at the outset soon changed into a light breeze and then into nothing at all. Ten miles from our final destination, there wasn't a breath of air. The surface of the ocean now looked like a huge convex mirror. A barely visible swell, making it undulate grotesquely, distorted the shapes of the few scattered clouds reflected on the surface. One might imagine for a moment, under this vast mirage, that a giant was dozing flat on his stomach and that his back was rising and falling in rhythm with his breathing.

The reassuring sensation of sailing on the back of a drowsing creature soothed my memories of wild seas. The stormy crossing of the Bay of Biscay was now nothing but a distant nightmare. This beautiful morning offered us its calm, and we had to take advantage of it. The barely blue sea made us feel we were in the midst of a fairy tale. Once upon a time, a prince from Vendée . . .

For a good fifteen minutes, far behind us, a stem wave that we thought belonged to a fishing boat had been rapidly approaching us. A few minutes later, a look through the binoculars removed any doubt; we were being followed by a ship of the American Coast Guard. When it was barely two cable lengths away, the siren went off. The loud signal was persuasive enough to make us immediately cut our engines.

When it came alongside, an armed soldier with a bull-horn on the forward deck spoke to us.

"We want to come on board to check your papers. Prepare your vessel for boarding."

"No problem, sir. Welcome aboard."

The huge Coast Guard vessel, whose forward section was painted with two red stripes, came up to our starboard side, and two men jumped on deck. I'd had enough time to prepare all our papers, including our authorization to be in the area. Our exchange was cordial, but with only the words necessary to impose respect, conversation was short and precise.

"Last-minute change, sir. The area is now closed to all sailing."

"But, sir, I have all the necessary papers."

"Yes, sir, you must leave the area immediately and sail off. Sorry, those are my orders. If you continue your current course, we will be obliged to take you to our base. You were picked up on our radar."

"But, sir, what about our authorization?"

"Sir, after the recent demonstrations on Vieques, we are taking no chances, the area is closed. Bon voyage and good sailing."

It was just one more incident. Once the launch had moved off, we steered a little to starboard to head away from the coast and toward Saint Croix. The sea was still just as fine. We would return to Salina del Sur Bay another time, since fate had so decreed. As for the spawning ground, we had the precise coordinates so that when things had calmed down we could set our hydrophones down there sometime in the future.

For now, we had forty miles to cover to reach the first of the Virgin Islands that Columbus had discovered. Six hours of sailing would be enough to get there, but the heights of the island would be visible long before. The bombs in the region had in the end given rise to a dialogue of the deaf.

I had only one idea in mind — to listen to the Rachmaninoff concerto again.

9

Two Accursed Destinations

The expedition charters a plane — Arrival of the researchers — Zandy and his signs — Sailing south — An accursed destination — The night of storms — Forty hours at the tiller — Night arrival at Nevis — A healing rest — Sailing to Montserrat — A sad visit with James — Shooting the film — A brief halt at Les Saintes — Involuntary vacation — A museum in the fort — A deep anchorage — Some thieves — Memories come back — Crossing the channel.

One never grows weary of glimpsing mountaintops on a new island coming over the horizon. "Unending flight/Intoxicating variety." Even if the sensation is already familiar to you, the same wonder seizes you every time. So it was for us on that late morning because, coming from the northwest, we would have a panoramic view of the entire width of Saint Croix, the largest of the Virgin Islands. Its jagged outline, barely coming over the blue horizon, already revealed its splendid and majestic beauty. Its thin coat of green rose gently from the liquid mirror, shaping its gracious form as though it were a green veil covering the body of a modest young woman coming out of her bath.

Columbus, who saw the island in profile coming from the east, was not immediately aware of its size. It was only when he rounded the point at the eastern end that he was able to grasp the length of the north coast and the large number of its sheltered harbors. Nor did he have what we did, after we changed course off Vieques: its long shape outlined on the electronic map of the navigation screen.

On our course we had just sailed over a depth of thirty-five hundred meters. This first stage of a four thousand meter-deep fault that runs along the southern coast of the island reveals the many sliding movements between the Caribbean and North American tectonic plates. Saint Croix is just on the edge of the fault, like a promontory overlooking from 4,350 meters a vast underwater plain extending south to the spurs of the mountain range on top of which is set the delicate island of Las Aves. This little island, our next destination, was the focus of all our preparations while we were on Saint Croix.

The Arawak name for Saint Croix was Ay Ay. We were now sailing just off Salt Bay. This was where Columbus dropped anchor, went on shore, and found at the end of a deep and narrow indentation a village from which the Indians fled when they saw him and that his sailors pillaged, taking "everything we pleased."

This was clearly a bad beginning, and as we know, things went from bad to worse until Columbus left the island.

To enter the port of Christiansted, you must not miss

the first buoy of the channel opposite Barracuda Reef or you might be headed for disaster. This long stretch, marked off with buoys, goes around the many coral outcroppings protecting the harbor.

When we tied up at the dock below Fort Christian, built in 1749, we were on a stage set for an operetta, with rows of pastel façades and shaded arcades.

Our arrival did not go unnoticed. The local television station had sent its star reporter, who was waiting for us, camera on his shoulder. We soon learned that an associate of the University of Miami had arranged a little celebration for our arrival. The *Prince* had been spotted as it was approaching the island. We were the people who were going to sail from their island with major researchers to an unknown spot to install bizarre instruments that would be used to predict earthquakes. After correcting some details, we appeared on the front page of the local newspaper with a more realistic headline: "Measuring the Advance of the Caribbean Plate — Beginning to Learn about Future Volcanic Eruptions." This media exposure helped us enormously in preparing for the expedition to Las Aves.

We chose a good mooring near the town and a marina's shipyards. Our dinghy served as a taxi to transport the small things necessary.

We had a week to find and load everything we needed before the arrival of the American researchers. The list was long. After locating the stores where we could get the tools, we had urgently to worry about whether Las Aves was still above water, because we were not absolutely sure. All the information we had gathered in Miami demonstrated that it still existed, but we had to verify it more thoroughly before leaving. The crossing by boat was long, and it would be pointless to go there for nothing.

Zandy, whom we had met on our arrival, worked for the television station. He and his wife Donna, who owned a real estate agency in the port, were a perfect couple. They adored each other and both of them had unlimited love for their island. Donna put her offices at our disposal for our e-mails, while Zandy, while he was not out filming, drove us everywhere we wanted to go in his van.

Zandy, who knew everyone on Saint Croix, introduced us to his friend Bob, a pilot who owned a small tourist airplane in which we flew, as mentioned, over Las Aves. However, what I haven't told you is that Bob had greatly underestimated the fuel consumption for the flight in case of unfavorable winds. The distance from the Saint Croix airport to the island was 120 miles, that is, 240 flight miles, plus an overflight of ten minutes. That corresponded to precisely a full tank of fuel, but not more. Winds at high altitude, great consumers of energy, were responsible for an unpowered landing on our return. Bob, fortunately for us, was a good pilot, and now that we had photographic evidence, we could prepare to load the boat for sailing to the island. Flying over the narrow strip of sand, we had detected no sign of life on the beach or near the containers. The island was indeed abandoned, or so we thought.

Zandy pointed out the best store to buy what we needed to install our satellite measuring instruments: two bags of cement, a pick and shovel, four iron bars, sixty liters of springwater, two large plywood boards, two twelve-volt batteries — everything on the list that Professor Dixon had given me during our last meeting in Miami. Nothing had been forgotten.

The *Prince* was also the focus of our attention. Hulls were brushed, diesel tanks filled, engines checked, oil changed, spark plugs, electricity, rigging, solar panels, on-board radio — everything was minutely examined. We would be on our own for at least two weeks, far from everything, with no possibility of immediate help.

When everything was finished, we had two days until the specialists' scheduled arrival.

Zandy, who was always ready to take a day off, suggested that we explore a site that only he knew about. He was, in the end, a true artist, full of imagination and resources. He had recently begun an archaeological dig on the southern coast of the island. Fascinated by the history of Arawak civilization, he had unearthed some strange shapes that he wanted to show us. Zandy must have been between forty and forty-five, with sharp features, an Indiana Jones fedora, and a brilliant smile. He looked like a tough guy, but he had a soft heart. He was enthusiastic about everything. If he had been free he would have come with us, but for now he wanted us to share his discovery.

In his van on the road along the southern coast constantly swept by the wind, I asked him if Saint Croix had experienced similar violence to that on Saint John.

"In 1755," he replied, "there were 375 French- and English-owned plantations here, growing sugarcane, cotton, indigo, and tobacco. The black slaves, better treated than on Saint John, remained quiet until 1848. The Danish governor at the time refused to abolish slavery, which led to a revolt, and the town of Christiansted was set ablaze. That was the beginning of the end for the plantations. Now we store oil and refine it a little farther down the road."

"That provides the island's livelihood today?"

"Yes, but not only that. We also have some tourist boats. But most important are the rich American families who have built magnificent houses here and spend the winter, like the Kennedys, for example. They appreciate the quiet and the beaches on the north coast. It's ecological tourism."

"Are there still vestiges of the Arawak on the island?"

"We're getting there."

He parked the van in a deserted spot above a cliff and asked us to follow him. Going down about ten meters, Zandy pointed out on one of the rocks overlooking the sea a little mark carved into the rock that was being worn away by the wind. There was another, even more clearly visible, farther on. The late afternoon sun, low on the horizon, allowed us to clearly distinguish the shape of a face coarsely sculpted on the rock. You could pick out the two eye sockets, the nose, and the mouth of this expressionless stone face. It looked to me like the same forms that we had discovered in the cliff caves on Barbuda, which was at the same latitude and a hundred miles away.

"I think," said Zandy, "that these signs indicated to other tribes the limits of the hunting territory onto which they were not allowed to go."

"A private hunting ground, so to speak?"

"Yes, that's it."

"And the date?"

"I sent several photographs to an archaeologist friend in the United States, and he tells me that these petroglyphs are more than three thousand years old."

"So, men and women lived here well before the arrival of the Arawak?"

"It's very likely, and what you tell me about your discoveries in Barbuda proves that they sailed from island to island, unless the continental shelf was larger then than it is now. But as for earlier than thirty thousand years ago I have doubts. It's possible that humans lived here before the migration across the Bering Strait that peopled the North American continent, but I wouldn't risk stating that as fact."

I left Zandy to his thoughts. He had just contradicted all the theories of the probable history of the peopling of the continent. But one can always dream. It's by paddling upstream that one comes to understand the river.

The next morning we went to the airport to meet Alberto, a geophysics student arriving from Chicago. He had brought in his luggage some of the measuring equipment that Professor Dixon had given to him. Three aluminum chests contained the satellite dishes, the computer, and the solar panels. Alberto, for whom this was not the first expedition, quickly went over everything that we had already bought and found everything in good order. He had brought a message from Professor Dixon wishing us good luck and setting a meeting for three weeks later on the island of Dominica.

Andy, Professor Dixon's colleague from Miami, arrived the next day from Mexico, where he had just measured the velocity of displacement of the Pacific plate in Baja

California. His plane, which was very late, arrived in the middle of the night without his luggage. He settled in on board as well, waiting for his four aluminum chests, one of which contained a drill. Fortunately they arrived the following day.

The first meeting with our two new friends went off very well. Intelligent and dynamic, they quickly adapted to life on board, although they had never sailed before. They were both eager to get to Las Aves to install their measuring instruments. Everything was brought and stored on board. Equipment was everywhere: in the hold, under the beds, in the engine compartment, in all the places where we could easily locate it later. If only the *Prince* had been roomier. I calculated that we had taken on more than a ton of equipment.

After a phone call to David, the meteorologist for the area on Tortola, to make sure we would have good weather for the crossing, we sailed at noon from the Christiansted marina. Zandy, his wife, and other friends saw us off from the dock. The Coast Guard had been informed, we were in the middle of an area of high pressure — what could possibly go wrong? The weather was magnificent, we left behind some good friends, and were sailing with new ones. The *Prince*, delighted to be back in the open sea, kept his peace, even though the waterline was a little lower than usual. We were 140 miles from our sandbank. At an average speed of six knots, we should arrive the following morning.

In order to head south, we first had to round the eastern point of the island, following the channel dividing Buck Island from the huge coral reef off the northeast coast of Saint Croix. A fair wind got us through quickly. To port, we left the Buck Island Reef National Monument that Zandy had taken us to see on the third day there. It is a true coral paradise, with clearly marked underwater trails. At the farthest end of the island, before it turns to the south, NASA has installed huge parabolic antennas pointing upward to listen to outer space. The program is not new. Any sign of life in the universe, any emission of sounds or strange frequencies, is analyzed here. The results so far have been far from encouraging. The program, whose budget had been reduced, I had been told at Cape Canaveral, continued only here. The researchers with their huge ears listened to the sky. More modestly, we were listening to the sea, or at least the fish. In any event, there were sounds beneath the surface and we had made some progress.

We rounded the point and headed out into the open sea, steering southeast at a heading of 165, with fairly calm seas, all sails unfurled, finally under way.

Our new scientific crew had quickly settled in on board. Alberto and Andy were already in bathing suits on the deck. It was steaming hot. Alberto told us than when he left Chicago the temperature was 36 degrees Fahrenheit. Everyone was delighted to take advantage of these few days of good weather to get a tan.

That afternoon we had our first work meeting. A detailed plan for our activities on Las Aves was put together. Unloading, surveying, installation of the satellite dishes, setting them to work — everyone now had his assigned duties. The trip also allowed us to get better acquainted. Andy had wide experience in measurement by satellite of the movement of tectonic plates. He had participated in the program that had installed this kind of instrument on several

Caribbean islands; they had to be set at regular intervals to measure the displacement. He and his team had been responsible for determining the annual rate of displacement of the islands in relation to the North American plate. The islands are caught in the vise created by the pressure of the contrary movements of the two plates, the Caribbean and the North American, provoking significant subterranean friction that gives rise to widespread earthquakes. Not only are huge underwater faults being created, but if the rate of displacement were to increase, the risk of terrestrial movements would grow. These studies are therefore of great importance.

The island of Las Aves had a huge advantage compared to measurements taken on more northerly islands: namely, it was right in the middle of the Caribbean plate. Measurements at regular intervals would lead not only to determining the risk of earthquakes farther north but also to pinpointing more closely the risks of volcanic eruptions to the east, where the Atlantic plate slides under the Caribbean plate.

Andy knew his subject well and, listening to him, our expedition, with this third area of research, became even more exciting.

But we weren't there yet. For the last hour the wind had been slightly strengthening. We had to reef in the mainsail, but nothing serious. I thought that the wind would slacken with approaching nightfall. David had said it would.

After dinner, we had to take a further reef. The sea changed from relative calm to agitation. The wind had veered to the southeast and disrupted our comfort on board. But the needle on the barometer had not budged. I decided to stay at the tiller during the night. Around ten, we had to take in all sails, the jib and the mainsail, and start our engines. At midnight the sea was already very heavy. I silently cursed David the meteorologist, who, it was claimed, was never wrong. What was going on? Our friends Andy and Alberto were showing grave symptoms of seasickness. At two in the morning, the boat was more and more difficult to control. Its added weight made immediate reaction at the tiller no longer possible. Our guests, now green, were lying on the bench in the wardroom. They no longer dared go down to their cabins. One hour later the sea was raging. Winds of fifty knots were howling aloft. I had been drenched for a long time. The spray was flying above the wardroom. It was pitch dark. The breaking waves were more than fifteen feet high. The *Prince* had started a uncontrollable dance, one that was incomprehensible for this time of year. The crew was no longer anxious, they all seemed to be dead in the hold of the boat, or, I suspected, wished they were. Unfortunately, I could do nothing to help. Around five in the morning we were in a full-fledged storm. A decision had to be made.

I had no idea how long this raging storm would last. Still, given the current conditions and the height of the waves, the island of Las Aves must be completely submerged and thus impossible to locate. Our position at the time was 16°30' north and 63°55' west. We had covered sixty miles, which meant we had another seventy to go. At our current speed, we could not hope to arrive before the following night. I was ready to take the risk, but I was not alone on board. The nearest shore was eighty miles away. There, at least, we would be sure to find calm, because Nevis is well protected on its west side. As steadily as I could, after setting the automatic pilot, I went into the wardroom to explain the choice we had to make. Andy, the older and more responsible of the two, listened to me between two bouts of nausea. I explained the situation to him. His condition scared me a little, but his mind was still functioning. He thought we could not take further risks, since we were uncertain whether or not we could land on Las Aves.

Our only immediate goal was to reach Nevis as soon as possible, at which point we would decide what to do next.

Once the decision was made, I set a heading of 80 for Nevis. Eighty miles at a speed of three knots meant that we would be there in roughly twenty-five hours. We would arrive at night, but it didn't matter because I knew the area well. Twenty-five plus sixteen; that would mean I'd be at the wheel for forty-one hours. Let's go!

Throughout the time it took to get to Nevis we had the same rotten weather. The *Prince* had been changed into a ghost ship. The friends had still not gotten up, apparently hoping they would die. It was only when we came to the leeward side of Nevis, and with the coffee that I could finally make, that they cautiously opened their eyes. Their nightmare was almost at an end, as was mine. Arrival at night is never advisable in the Caribbean, but we had no choice.

It was strange to know that only three miles away loomed a huge volcano more than three thousand feet high, which could have guided our approach to the island, but the night was so dark that we couldn't see our hands in front of our faces. I could only detect a few points of light ahead to port that must have been those of the little town of Basseterre on Saint Kitts, and to starboard the lights of Charlestown, the capital of Nevis. Relying on my memory and the map, I decided to sail down the middle of the channel dividing the two islands, and then to steer to starboard when even with the lights of Charlestown and drop anchor in the large bay north of the capital, which is quite wide and not encumbered by any shallows. It was four in the morning. There was not a ripple or a breath of wind, the silence was total. In the dark, I dropped anchor at a distance that I guessed was far enough from shore. We had finally arrived. The crew, surprised by the calm of the spot, began to open their eyes. A few smiles reassured me about their condition. Andy, no doubt in the best shape, joined me, staggering on the rear deck, to take advantage of the cool night air and fully recover his spirits.

"You okay?" I asked.

"I'm feeling better."

"Good."

I couldn't ask him for more. I only had one thought in mind — sleep, sleep, and more sleep.

The good smells of grilled meat and sautéed potatoes woke me up. It was noon, and everyone was famished.

As I emerged from my cabin, the cone of the volcano loomed directly above us. The *Prince* was half a cable from a long beach, which was bordered by immense coconut trees. Our mooring was perfect.

While I was sleeping, the crew, totally recovered from their two nights of horror, had prepared a superb meal. It was, they told me, a way for them to thank me. For the occasion, I took out our best bottle of old Bacardi rum. We wouldn't shed any tears about our luck: we were alive, weren't we?

It was also Columbus who had discovered this island on his second voyage. He found it so high that he called it Las Nieves, "Snows" in Spanish. For me, Nevis is the most beautiful of all the Caribbean islands. Not very well known, seldom visited, except by a few rich English retirees, it still has the charm of the West Indies of the past. Politically joined to the island of Saint Kitts, it is easy to imagine in its harbor a good dozen ships of the line at anchor and two or three square-masted brigs. This queen of the Caribbean was the birthplace of Alexander Hamilton and had provided shelter for Admiral Nelson's amours. The panorama created by the little huts perched on the slopes of the volcano covered with a dense forest is a delight for the best primitivist painters. Here, too, five years earlier, I had been extremely lucky to find an Arawak archaeological site. The few pottery shards I had unearthed had been deposited in the little museum in the capital.

Because of what had happened, we were ahead of schedule. Once again, the sea had decided for us. The local fishermen told us that they had never seen such weather at this time of year, and they did not anticipate any improvement in the coming days.

"It's only a postponement. We'll get to the island one way or another. Don't be disappointed," Andy said.

"We'll need a larger boat, and we'll sail from Guadeloupe," I replied.

"In any case, we're expected on Dominica in twelve days. And we have all the equipment on board that we can use there. I'll take the plane there so I'll have time to prepare."

"Okay, I'll keep Alberto on board; we'll raise anchor tomorrow. I'll stop off on the way at Montserrat. We'll see you in twelve days. We'll be anchored in the bay on the northeast of the island, just opposite Portsmouth. We'll unload all the equipment there to begin work on volcanoes with the other team that's joining us." I also thought to myself that Andy wouldn't be all that anxious to set foot on another boat.

Exactly twelve days later we would indeed be there, and we had a little time to spare.

In the afternoon, after driving Andy to the little airport on Nevis in Jack's car, we all took a tour of the island. The vegetation, fed by the many streams coming down from the top of the volcano, was extraordinarily luxuriant. The road, sometimes overlooking the open sea, sometimes the leeward shore, offers a variety of splendid vistas, particularly on the narrow arm of the sea that divides the island from Saint Kitts. In the distance, we could see the island of Antigua. The open sea was still raging.

The next day, when we had barely sailed past the southern tip of Nevis, we could already see the island of Montserrat. Huge clouds of brown smoke were spewing from the dome of its volcano. La Soufrière had not spoken its last word. With our luck, we had better expect the worst. The sea had already come down on us, and now perhaps the sky would fall. Our crossing was agitated but brief.

A flimsy craft anchored not far from us at Nevis sailed out at the same time. On board were an English couple completing a tour of the islands. They were sailing to Trinidad to store their boat in safety for hurricane season. The *Prince*, much faster despite its load, soon left them behind. They reached Montserrat three hours later for a brief overnight stay.

The only bay where mooring is allowed is located in the northern part of the island, which is where we intended to shoot our second short film. Alberto, who had a local volcanologist correspondent here, arranged for us to meet.

Facing the bay, which was bordered on either side by steep cliffs and hidden by old containers, were the offices for securing entry permits. Waiting for us there was James, the volcanologist, whom we had contacted by radio during the crossing. We agreed that he would pick us up in his jeep early the next morning. Meanwhile, he would take care of authorizations for us to cross the fence dividing the island in two. Since 1995, when hot volcanic ash had covered the southern part of the island, it was barred to everyone but the police and scientific researchers.

James was English, but had studied in the geotectonics department at the University of Chicago. He was in charge of surveillance of the Soufrière volcano. The island sun had had no effect on his red-spotted face, which retained the pallor of his native land. Stoic and full of humor, he was an invaluable guide for our sad tour of the island.

On the other side of Belham River, whose bed was choked with huge blocks of lava, there was no vegetation left. Even at a distance of six kilometers from the summit, the heat from the ashes had burned everything. The few tree trunks left standing lifted their scorched branches skyward, as though begging for mercy. On the valley's sides, however, a few tufts of green were beginning to show. The road was gone. The asphalt had melted, and what remained of the roadbed was blocked by large chunks of gray stone. We were no longer in the lush Caribbean but careening through a lunar landscape in which the rich colors of the lost vegetation had been replaced by gray and brown. Our cameras made a permanent record of this apocalypse.

A little farther on, a long barrier of mixed fencing and barbed wire cut across what remained of the road, dividing the island into the two zones. Three armed police officers guarded the entry. They searched the jeep and asked for our papers and our permit. They also told us that we had to return to that spot in no more than four hours.

From atop the next hill, we could observe the extent of the disaster. The thick mass of the volcano formed the backdrop of this dark picture, the dark plumes of smoke that still drifted from its mouth shadowing a landscape that had taken on a steel gray hue. One could only guess at the presence of what had once been the little town of Plymouth. Only a few gray roofs and scorched spires emerged from an immense carpet of ash. There was nothing left, except that one could still see to the right, on the shore, the arms of two deserted jetties. A deathly silence pervaded this lifeless space. I had trouble mastering my emotions. No birdsong or cock's crow troubled this vision of sorrow. Life had stopped here six years before.

James, as affected as we were by the dismal scene, even though this was not his first visit, spoke in a whisper in order to respect the silence:

"Seven thousand people were evacuated from here. In late 1995, they had to leave behind everything: their hospitals, their schools, their houses, their possessions. Few of them stayed on the island. England and Antigua took in many. Fortunately, there were very few deaths. This zone is now totally off-limits because of looters. Since then, the volcano has never stopped spewing ash. A surveillance system has been set up. We've installed a number of sensors on its slopes which do nothing but record the volcano's activity. Sometimes the slopes are again covered by flows of ash. It is still rumbling. Its smoke almost constantly darkens the sky above what remains of Plymouth."

We went down to the edge of the town. Thick walls of ash hid from view what remained of streets and houses. This desert of gray, hilly dunes had buried everything. Here and there were a few burned-out cars, all without tires, the only sign of past human activity. The red and white circle of a sign indicating a one-way street rose above a hill of volcanic dust. A piece of the leg of a plastic doll also emerged from the ash. A slight odor of sulfur drifted through the humid air. That was all.

James went on, still whispering: "We can't get near the summit; it's too dangerous, because there have been a few tremors in the last week, and our instruments indicate an imminent seismic disturbance."

"The two tectonic plates are not about to stop sliding under one another," I answered.

"That's true. We need more frequent measurements of their velocity to see things more clearly. The more the velocity accelerates, the more likely it is that magmas will rise to the surface."

"We tried to go to Las Aves to measure the velocity of the Caribbean plate, but bad weather kept us back."

"A pity."

"You're telling me! But we'll try again in the very near future."

"It will always be tough to predict eruptions, but every scrap of information helps."

After what we had seen, we returned to the fence in silence. We were not about to forget that vision.

As we passed the police barrier, one of them said to James, "With all your wonderful technology, there's no way you can predict such eruptions."

James replied with a smile and a friendly wave, then said to me when we had proceeded a little farther, "They think it's easy. We're just at the beginning of the science of volcanology. We still have everything to learn. We're a long way from making it an exact science."

Professor Dixon had told me much the same thing.

We left the island the next morning at dawn, after inviting James on board the night before. He had heard about Professor Dixon's work. When I told him that Dixon would be meeting us on Dominica for some work on volcanoes, he asked us to send his greetings. He said he would send him an e-mail with some observations he had just made of the volcano. International scientific solidarity was a reality, which was all to the good.

Two police boats crossed each other just off Plymouth. Seen from the ocean, the volcano seemed even more threatening. The shadow of its clouds of dark smoke was carried far out to sea.

The more cheerful Guadeloupe was already in view. We were only thirty miles away. Sailing in the calm waters of its leeward coast, we passed in the distance its volcano, also called La Soufrière. Its summit was concealed behind a thick layer of clouds. Kept under very close watch by all kinds of machines hooked up to computers, it had recently witnessed a dispute among major theorists of volcanology, each of whom had a different opinion about the date of its next eruption. This active volcano is part of a group containing eight others located on the Caribbean arc. They are all the result of the sliding of the two oceanic plates over one another at a very great depth.

Before we arrived in Dominica, I wanted to get a shot of all the Îles des Saintes to complete my film. When we crossed the channel separating these islands from Guadeloupe, we were greeted by strong contrary winds and a violent current — which made me wonder if the reason the sea was so often bad off Las Aves was that all the currents that circulate in all the channels dividing the Caribbean islands converged at that spot.

From atop the ramparts of Fort Napoleon overlooking the Baie des Saintes, the view is magnificent. You can clearly discern the outlines of the collapse of the summit of a volcano that must have existed about 300,000 years ago. A few domes still exist, such as the little isle of Cabris and the chalky Morne encircling the crater.

But the happy surprise was to find in the fort a collection of very realistic paintings tracing the life of the first Indians on the islands. There are a dozen affixed to the walls of an inner room of this very well maintained fort. The artist, whose name escapes me, clearly had a lively imagination. Everything I had seen came back to me. The cliffs of Barbuda, the windward coast of Saint Croix, the site of Nevis — all made sense. Even if the paintings were probably not that close to reality, the representation of those human faces was a no doubt ephemeral proof of their past existence.

After recovering, with some difficulty, the motor of our dinghy, which had been stolen, it was time to rejoin our team on Dominica. The *Prince* threaded his way around the many small islands of the archipelago to reach the channel that was to lead us to our next stage. Once past the large island, we had a clear view of the volcanic peaks of Dominica. For once, the sea was as smooth as oil. We would be on time on the appointed day for our meeting with the rest of the researchers.

It was time to listen to a little more Rachmaninoff, which Alberto and Raphael now shared with me.

10

Satellites and Volcanoes

The lemonade of the battle of Les Saintes — Flora — The fourteen volcanoes — A tricky installation — Meeting the second team — Looking for the right spots — The installation of satellite dishes — Geologists at work — First measurements — Night work — Authorizations — Lunar conversations — Priests from Vendée — Martinique sheltered from hurricanes — The Prince *in safety.*

We sailed slowly toward the wild and beautiful island of Dominica. At its base, its exact reflection was spread on the mirroring surface of the sea. This inverted image of its mountains barely crowned by the morning fog perhaps still contained the distant memory of a tumultuous past. For it was in these waters in 1782 that the famous battle of Les Saintes had taken place.

"Make me a lemonade," Admiral Rodney ordered a young midshipman, tossing him the lemon he had been sucking (close-up); the midshipman returned a moment later, stirring the bitter drink with a dirty knife, the only utensil he had been able to find. The admiral looked at the knife and said, "My boy, that might be good enough for a midshipman, but not for an admiral. Drink it yourself, and send me my steward."

Standing on the poop deck of his ship, the *Formidable* (tracking shot), he had just cut through the line of French vessels and knew that victory was within reach. His new tactic had totally surprised the enemy fleet. He could calmly drink his lemonade. I was already imagining the shots in the film recording the events.

Bougainville was fifty-three. He was head of one of the squadrons in the French fleet commanded by Admiral de Grasse. After his discoveries in the Pacific that had caused such a stir among his countrymen, he had returned to service in the navy. A change in the wind at the last moment had dispersed the thirty-one French vessels. It had suddenly veered to the southeast and forced them to take in their sails. The maneuvers, which could not all be done at the same time, created gaps between the ships, into which the British sailed. This new tactic put an end to the current practice of the line of battle. It was revenge for the battle of Chesapeake Bay. The British had become the new masters of the South Seas.

Friction sprang up between Bougainville and Admiral de Grasse when they returned to France. Each accused the other of responsibility for the defeat. Bougainville, who had strictly obeyed the admiral's signals on his ship, the *Auguste*, was acquitted by the court-martial held in Lorient. Disappointed by Principal Minister Loménie de Brienne's rejection of a proposed expedition to the South Pole, he nevertheless supported Louis XVI to the bitter end, upon which he was thrown into prison. Liberated through the bravery of his wife, he later became a protégé of Napoleon and died at the height of his fame at the age of eighty-two at his desk, next to which he had set a necklace of iridescent shells he had brought back from Tahiti.

We were now on the lee side of Dominica. Two fishermen off Prince Rupert Peninsula waved to us in welcome. Their boat, painted red and green and equipped with an antiquated outboard motor, came alongside. Wearing old tattered T-shirts and standing barefoot on planks covered in seawater that they had not yet bailed out, they spoke to us with strong English accents.

"Fish for you. We've just caught it."

"Okay, we'll take some. How much?"

After we picked out two fine groupers and a porgy, they went on: "We'll go with you to the other side; we know a good mooring for you off Portsmouth."

One of them came on board.

Isaac was his name, and he had eyes as piercing as salt crystals. Ageless, his face masked by a shaggy beard, he moved with measured gestures. His broad, fixed smile indicated that he was very happy with his modest condition as a fisherman. The luxurious *Prince* in no way hampered his friendly and easygoing manner. He was at home on board. Drinking his coffee and eating the bread we had offered, he curiously inspected all the instruments on board, and asked, "You're working for science?"

"Yes, that's right."

"And you're interested in fish? Because I know the good spots."

It was a shame; Isaac didn't know that we had come here not to listen to fish, but to work on volcanoes. Otherwise, he could have become a loyal research associate, like Joseph, John, and Moose on Barbuda. For a moment, I recalled my adventures on the north lagoon of Barbuda. There must have been good spots on Dominica as well.

Isaac went on: "My brother Zach puts out nets out at sea. When we go there once a month, there are huge fish around them. That's where you should go."

I had already heard about this method from a fisherman on Saint Barts. You had to go far out to sea and then dive to identify the species lurking around the heavy net. Using iron pipes as ballast, the broad mesh harbors many small organisms that are food for marine fauna. All this was not free of risk. It seems that there were a large number of sharks usually swimming around the tuna to feed. Some dolphinfish, drawn by the small species, also used it as a favorite hunting ground. This ocean larder, a kind of trap for the relentless cycle of life, was a living example of the food chain. It would certainly have been a good spot for listening, but for now we had to concentrate on the volcanoes of Dominica. And that would be no small matter!

Isaac led us toward an imposing concrete jetty at the southern end of Portsmouth Bay, where a number of old tramp steamers were already docked. Indispensable links for exchanges among islands, most of them were badly maintained and covered with rust. One of them was unloading sacks of flour from Trinidad. A crane, whose fragile arm rotated on the platform with a high-pitched squeak, one by one deposited torn sacks on the backs of the Guyanese crew. In spots, the jetty had become white, like the faces of the men bent over from their burdens. It looked like snow in the month of June. The graceful schooners had long since vanished.

The Customs and Immigration Office was just at the end of the dock. With the help of our new friend Isaac, all the forms were soon filled out.

At the north end of the bay, Devil Hill, covered with dense wild vegetation, overlooked the broad expanse of the deserted bay. Farther south, Indian River, nestled under a thick canopy of coconut trees, creepers, and giant roots entwined together, quietly discharged its clear water onto the gray sand of a minuscule beach. We had without a doubt finally arrived on Dominica.

Isaac departed after making us promise to come to see him at nightfall.

Andy, the researcher who had left us when we were on Nevis, had spotted the *Prince* from the road. We rounded Point Bluff and tied up at the dock, and he arrived a few minutes later.

"Welcome to Dominica, and congratulations for the date. You're right on time. How was the sea?" He had not forgotten his turbulent crossing.

"Thanks for the welcome! A perfect sea, nothing but good weather. What's the plan?" I asked.

"I've come with a driver. We have two Japanese four-wheel drives parked behind the building. If you agree, we can unload everything this afternoon."

After lunch, which we all ate on board, Andy explained that the rest of the team would fly in the next evening. He had decided to establish working headquarters in a little hotel in Roseau, the island's capital. He would transport all the equipment there to check that it was working in the late afternoon. He would wait for us the next evening on the pontoon opposite the hotel. He had picked out a buoy on the left side of the bay to which we could tie up the *Prince*, and he invited us to stay at the hotel.

"It will be easier, because with what we have to do, you never know what time we'll get back from an expedition," he explained.

It took us two full hours to unload the boat and transfer all the equipment into the trunks of the two four-wheel-drive vehicles. When they were loaded to the absolute limit, Alberto, the driver, and Andy set out for Roseau. Thirty kilometers of a very winding but very picturesque road awaited them on their way to the capital. The *Prince*, which had recovered his original waterline, was straining to get under way. But he would have to be patient.

Isaac came to get us just at the moment when the sky was blazing bright red. The sunset that night, like a talented director, had inflamed the circle of neighboring mountains. It had for just an instant stained the tiny hovels perched on the mountainsides with the colors of houses in Hollywood.

Accompanied by his dog, Isaac had come in his pickup truck. Neither the make of his truck nor the breed of his dog was recognizable — the truck was patched up with an incredible number of spare parts, and the dog was the product of countless generations of cross-breeding. Isaac lived in the little town of Toukari, north of Portsmouth. We brought along two bottles of good wine, the cheese and ice cream we had left, and one of the few remaining T-shirts embroidered with the expedition logo.

His wife was waiting for us on the wooden steps of his modest house. Her name was Flora; she was much younger than Isaac, and was expecting a child. Her straight black hair surrounded an unadorned face enlivened by two large brown eyes. She must not have been born on the island.

The fish soup she had prepared had the delicious aftertaste of the spices she had added, a little ground anise, some cloves, red pepper, and a little lime juice enhancing the rather bland taste of the fish. With few resources, Flora had decorated her house nicely. Some lace curtains, a few pieces of white Formica furniture, and a four-poster bed painted blue gave the house the clean and simple air characteristic of the islands.

Originally from Suriname, she had met Isaac in unusual circumstances. Orphaned at a very young age and brought up by her sisters in Paramaribo, when she was of age she was sent to Trinidad to work in a restaurant owned by some cousins of her late mother. Located near the oil refineries, the restaurant didn't have the most distinguished clientele. Quickly trained and turned into a maid of all work, she became more than a docile waitress. Her time off was rented out against her will to passing men, and her silence was the price she paid for not being sent back to her homeland with no means of survival.

She escaped after a short time in the middle of the night and sought refuge with a fisherman friend who brought his fish to the restaurant every day. A friend of the fisherman's, who owned a little tramp freighter traveling among the islands, offered to take her on his next trip from Trinidad to Guadeloupe. He knew the owner of a hotel there who was looking for personnel and who certainly had work and lodging for her. A hurricane that destroyed a large part of Dominica put an end to this journey. Sailing off the island during the storm, the tramp steamer had lost its tiller and was wrecked on the shore. Isaac, who had seen what was happening, hurried up with his boat to save the crew. This was how he met Flora. From that time on they had been living together. As we were leaving, Flora said, "I hope to bring my last sister here; she's still in Suriname. I would like her to be with me to help take care of the new baby."

After thanking her for dinner, Isaac came with us on board.

"What a woman!"

"I know," answered Isaac. "Thank you for the T-shirt as well; I'll wear it to church on Sunday."

As he headed home, he waved, no doubt to wish us good luck.

We set sail early the next morning. Roseau was seventeen nautical miles away. Sailing along the coast on the smooth surface of the water protected from wind by the mountains, it would take us three hours to go and tie up at the buoy that Andy had found for the *Prince*. Seen from the ocean, Dominica resembles a long line of cathedrals of vegetation whose wooden arches plunge directly into the sea. Imp, Crown, and Three Peaks make up the principal spires, some of which are fourteen hundred meters above sea level.

For a century, the island had been an impregnable refuge for the Carib Indians, who had fiercely resisted the French and English invaders. Indeed, an exuberant humid forest covers the entire island. Now the Caribs are confined to a reservation of a mere two thousand hectares in the northeast.

The *Prince* glided like a swan on these calm waters. A few fishermen in their gum-tree canoes gave us friendly waves as we passed by. They were fishing directly on the bottom using lines with several hooks. Standing in their unstable flimsy boats, they gave us the feeling that we would have to rescue one or two of them and take them on board. But no, their skill had long ago triumphed over the laws of equilibrium. Their childlike simplicity was the trait they shared with the fishermen of Barbuda.

Roseau was now directly ahead of us. Behind the cement dock that doubtless served passing cruise ships, it was easy to pick out the buoy we were to tie up at. Once the *Prince* was safely moored, we could make out on the docks a row of dilapidated and mismatched façades. Part of the town had been set higher up on the slopes of Watt Hill. My first impression was that this little town, built in disorder, with no real thought taken for the future, was a constant risk for its inhabitants, with a volcano looming over it. It was set in a valley that, in the event of an eruption, would serve as a conduit for the lava, which would immediately destroy it.

I recalled what had happened on Montserrat.

The whole team had now arrived at the hotel: Tim

Dixon; Glen Mattioli, who had just been elected to the National Science Foundation in Washington; and Alan Smith, professor of geology at the University of California at San Bernardino, for whom this was his second visit to Dominica. This was distinguished company indeed: scholarly men, young and talented, who throughout our stay never hesitated to share their wide experience with us.

There were now six of us, counting Alberto and Andy.

"I'm very sorry about Las Aves," said Tim. "Andy explained the situation for me. You made the right decision; no new scientific information is worth risking lives."

Tim provided me with great comfort, for I had not yet swallowed the setback. I described our visit to Montserrat and the film we'd begun to shoot.

During our first dinner together, we decided to split up the work between two teams, each with its own four-wheel-drive car. Tim, Glen, and Alberto would explore the central part of the island and scout out good locations for installing satellite dishes on the slopes of the volcanoes, while the second group would do the same thing in the southern part of the island, focusing on installing dishes near the shore, where evidence of the movement of the Caribbean plate was more pronounced. But first we had to secure written authorization from the government, because in certain instances, for reasons of safety, we might have to install the dishes in either protected or inhabited zones.

The next day, I went with Tim and Glen to one of the administrative offices. The chargé d'affaires for the territory was expecting us. He had held the position for only a week and was busy with plans for roads and a hospital soon to be built above the city. Expressing goodwill, he was prepared to listen to us and to grant the official authorization that would facilitate our work, but first he wanted to get a clearer understanding of our purpose.

Tim jumped in: "The United States Defense Department has launched twenty-four satellites around the globe, making it possible in any place to determine the time, our position, and the speed at which we are moving on a special device that receives signals through a particular kind of antenna. In short, it provides latitude, longitude, height above sea level, and variations in them." Tim obviously did not mention in the conversation the complex automatic calculations made by the software that enabled the result to be obtained.

"How does that concern us?" the official asked.

"The use of this portable material now enables very precise measurement of movement of the ground. In the specific case of your island, we are going to measure possible deformation of the volcanoes and in other spots the velocity of displacement of the plate on which the island sits."

I think he understood the principle of these measurements: Tim's explanation had been simple enough.

"Yes, I see, but how is that of interest to our island?"

For Tim, that was obvious, but there was a piece missing in his explanation. "This system of measurement has a direct consequence for the prediction of future volcanic eruptions. The greater the deformation of the slopes of the volcanoes, the greater the risk. The same idea also applies to the velocity of the displacement of the plates — the greater the increase in velocity, the greater the risk."

Logically, the official concluded, "So, we'll know in advance when it's going to blow."

"We're in a phase of cross-checking information for the moment. This new science can provide ideas about dates, but we're still far from making exact predictions."

"It's a beginning, so to speak."

"Yes," answered Tim. "A promising beginning."

Our authorization would be ready early the next morning. We would simply have to come to his office to pick it up, and the work could then begin.

Carrying all our equipment, our team took the road south. We each had a satellite dish and all the tools necessary for its installation. Alan the geologist had a detailed map of the island's geological formations. He intended to double up his work by collecting charcoal that should be found trapped in the pyroclastic layers spread through the region.

Three conditions are necessary in selecting a spot to install a dish: a rock has to be firmly anchored in the ground, the spot has to be in a clearing to be able to receive satellite signals, and finally the site has to be secure, because the material has to be left for three days — we couldn't afford to risk having it stolen. Alan said, "Generally, people are afraid they might hurt themselves if they touch the dishes, but in Central America we had several batteries stolen."

The first spot chosen was at the southern extremity of the island, on a rocky promontory named Point Cachacrou. The three conditions were fulfilled. Three hours were enough to put everything in place. First we had to drill a hole in the rock in which was set a steel rod called a pin, then we needed to fix the pin firmly in the hole with fast-drying cement, wait a few minutes, and mount the pointed base of the dish in the notch at the top of the pin. The dish is held upright with the help of three sliding arms fixed in the ground. The dish had to be perfectly horizontal, so it took a while to adjust it with the help of a level. Finally, everything was connected with a special cable to a computer concealed in a plastic case and powered by a twelve-volt battery. All that remained was to program the computer with respect to the date, the time, and the location. It could then begin to record the movements of the ground in its memory. I was able to film all phases of the installation.

After we came down the steep slope, we informed the police in the village located at the foot of the promontory. Thanks to our government authorization, the police would make regular rounds to keep an eye on our installation.

That morning, we had been fortunate enough to have a magnificent view of the southern part of the island. A large portion of the chain of volcanoes was visible. Deep valleys cut through this green mass. At the foot of each volcano, at the edge of the shore, a fishing village was nestled along a gray sand beach. The promontory also overlooked a huge bay whose clear waters provided a glimpse of large coral reefs.

On the horizon, the island of Martinique just barely pierced the heat haze to unveil the summit of Mount Pelée. To the west the surface of the ocean was white with foam; the sea was raging. We hoped that the *Prince* would hold fast and not break loose from its moorings. It was lucky that Raphael had decided to stay on board that day.

The afternoon was entirely devoted to looking for another spot. While devouring our sandwiches in the four-wheel drive, we crossed through the rugged terrain of the southern part of the island on a winding road. Very high flows of volcanic ash had been shaped by bulldozers to build the road. Alan, who was always vigilant, looked carefully through the open window to see if he could locate any traces of charcoal in the cleared spaces. We often moved at a crawl to satisfy his curiosity.

He signaled to me to stop the vehicle on a curve, because he had caught sight of a fine black mark on one of the ash heaps. He had seen it clearly because he was used to it. Located seventy or seventy-five feet above the road, almost at the top of the heap, there was indeed a vein of charcoal. But it was impossible to scale the steep slope. The only way of getting samples was to dislodge them by throwing rocks at the spot. He set a large piece of clear plastic at the base and we began throwing. Alan, who had a good sense of humor, said, "You see, geology leads to everything, even throwing stones."

The second throw luckily deposited a large quantity of charcoal on the plastic. Some pieces were large enough for later carbon-14 dating. He carefully put all those pieces in a small plastic bag.

"Then we'll know the date of the last eruption," he said.

He went on: "During an eruption, volcanic flows carry along everything in their path, notably trees. Heat and flame transform the trees into charcoal. A simple carbon-14 analysis provides the date." When he had been here two years earlier to prepare a geological map, Alan had also brought back some samples. Analyses had indicated that the last eruption had taken place three thousand years ago. The experiment was repeated several times during the day. He could easily determine whether or not each flow came from the same eruption.

"Will this information enable us to predict future eruptions?" I asked.

"Yes and no," he answered. "Geology informs you about the past and tells you whether your ideas about the future might be right or not. The current method, using GPS devices, shows you what is happening today and enables you to make predictions about possible volcanic activity . . . But geology is not confined to the past; it can also give you possible scenarios for the future."

Little by little I was discovering a fascinating world of which my apprenticeship in Miami had given me only glimpses.

In the field, encircled by volcanoes, I found myself in the best university in the world and no doubt with the best professors. I realized how lucky the *Prince de Vendée* expedition had been to be able to persuade all these highly qualified researchers to take part.

Late in the afternoon, we located the second site for installing a GPS dish. It was at the foot of a cliff overlooking the sea on a rock large and solid enough for the satellite dish. The owner of a small hotel located on top of the cliff would keep watch over it. That would be our next destination in three days.

When we returned to our base late in the evening, I first made sure that the *Prince* had withstood the strong wind. It was still there. Raphael, like the good sailor he was, was sleeping on board.

The other team had also found one good spot to install

a satellite dish, but had not had enough time to find a second. The dish was firmly installed on the slopes of Three Peaks, where it measured the possible expansion of the volcano. The entire group was satisfied with this first day's work. The following days were taken up with approximately the same routine — finding good locations on the geological map of the island, taking down the dishes, installing them, making sure they were working, programming them, and exploring again, wandering among the many hills. Hours and hours in a four-wheel drive were necessary to accomplish our mission. Bouncing over often muddy tracks or on badly maintained steep roads in extremely humid heat was our daily program. We were doing the rough fieldwork of modern volcanology.

A few days later, the two teams met on the other side of the island to look at a place Glen had discovered. An old lava flow had reached the sea; swept by the surf, its long shape plunged into the ocean. The silica that was its principal element gave this giant finger the appearance of a fragment of a black glove forgotten by nature. The installation of a dish on this giant appendage had considerable advantages but also serious risks. The opinions of all the specialists were important in arriving at a decision.

Some said the measurements obtained here would give information on the movement of the plate and at the same time on movement created by internal pressure in the volcano, while others judged the risk too high, because we could lose all our equipment if the sea were to cover up the flow. Finally, because we had only three dishes, the plan was abandoned. It should also be said that each dish cost more than $150,000 — enough to give me pause!

In our conversations over dinner, Glen often brought up subjects out of the ordinary. The theory that the moon was irrevocably drifting away from earth he found particularly fascinating. What would happen to everyday life if its gravitational effect weakened? Tides? The length of days? The entire cycle of life on earth would be completely changed.

He explained that when Neil Armstrong had landed on the moon on July 21, 1969, he had put a mirror on the surface at the request of NASA. By measuring with lasers its distance from the earth, it was found that the moon was moving away at the rate of one centimeter per year. Imagine for a moment the consequences after one million, then three million years! You would have to consider the ocean masses, which by then would be changed because of continental drift, then the time the moon would take to complete an orbit, which would have grown larger and longer, and finally predict the climate changes due to the shift in the angle of its axis of rotation and the consequences of its moving closer to the sun. Would our biological rhythm be changed if our species survived its follies? Would days last forty-eight or sixty hours? Would seasons still exist? Would we have a winter lasting two years? Often contradictory opinions enlivened this dialogue of eccentric geniuses. The evenings flew by all too quickly.

Every day we set out again to install our dishes or move them to another spot. This was a healthy return to reality. In our frequent travels, we often had the pleasure of meeting the inhabitants, most of whom were farmers. We asked them either for access to their fields or to help us find a more protected spot. The rugged mountainous island terrain often caused problems. More than once we were helped out of a tricky situation by a passing tractor. Even though our four-wheel drives were supposed to be able to overcome all obstacles, we were often bogged down, for example, in the middle of one of the many rivers that flowed down the slopes of the volcanoes. Unable to extricate ourselves, we once spent more than half the day unable to move forward or back. The geological map that dictated our destination only infrequently matched the road map. After three weeks here, we ended up with a through knowledge of the island and its dangers.

Often, when we passed through hamlets in the south, a family on their doorstep would invite us to share the produce from their garden, or even offer local artisan products. Some of them were more worried, and asked whether their school or their clinic was at risk in its current location. In most cases, these buildings were set at a distance from the flows, often on a height.

In a little village of a dozen houses set on either side of a dirt road, nestled on the summit of one of the hills adjacent to the tallest volcano, an old gentlemen, who was walking with a cane, spoke to us:

"You must be the volcano listeners. Welcome to my beautiful island. You know we're comfortable here. No need of television to be happy, we have everything we need. We would rather have better roads to get to town, but we make do."

The conversation continued on the advantages and drawbacks of life in the country. Finally, he arrived at the conclusion that you could be happy anywhere.

"Are you going somewhere? We can take you if you like. Let us help you into the car."

He was going to see his son, who owned a field a little farther down the volcano slope, to give him advice on how properly to prune his pepper plants.

"And you know when it's going to explode?"

Everyone asked the same question; we did too.

"It's won't be tomorrow, don't worry." This was the only answer about which we were sure.

"One of you is French, I think. He has an accent I recognize. Is it you?"

"That's right," I answered. "A loyal native of the Vendée."

"Really, so was my great-grandfather. He came to visit his brother, who was a priest in the north, and decided to stay here and get married. You should visit the priests, they're all from your part of the world. A very friendly bunch!"

His love of the past and of nature, still part of his genes, clearly showed his origins in the Vendée. Indeed, as the *Prince* said, they're everywhere. His son was waiting for him at the edge of his field, where we bid him farewell.

A few days later, when we had decided to give ourselves a day of rest to think about plans for the ensuing days, I drove toward the northern part of the island. Raphael, who had not had the time to travel around the island as we had, accompanied me.

When we came near the village of Salibia after two hours on a winding and often dangerous road, I asked a farmer for directions to the priests' house.

"It's at the bottom of the hill, you'll see there's a river, and it's just after the bridge on the left, there's a little road, it's right there. You can't miss it," he said.

The spot exuded peace. Hidden from the road by two huge royal poincianas and a row of royal coconut trees, the house was in the middle of a garden of hibiscus. In back, up against the hill, were a few rusted crosses, probably the vestiges of a tiny old cemetery. To the right, a multicolored statue of the Virgin was set on a windowsill of a chapel half buried beneath the leaves of banana trees. A slightly dented van was parked in front of the well-kept house. The ground was strewn with red flower petals from the poincianas. Two priests, alerted by the noise of our car, emerged from the house and came to greet us with a smile. They both wore jackets with a small cross on the lapel.

"Welcome to our humble home. You come from France?"

"Yes, how did you know?"

"That's easy, only the French come to see us like you. But there aren't many here. Where in France?"

I answered, "From the Vendée." There was no point in telling them then that I had been living in the United States for twenty years.

"Really, from what village?"

"From Pouzauges, in the *bocage*."

"The wild woods, we remember."

From then on, it was as though we had the right passport — we could enter the house.

The older of the two had been on the island for twenty years, the younger had been there for little more than a year, having replaced Father Villeneuve, who had retired to the Vendée after more than forty years on the island. Both had gone to seminary in Chavagnes-en-Paillers and both had been born in the Vendée. Members of the Congregation of the Sons of the Immaculate Virgin, they resembled all the priests from the Vendée serving mass in the islands and around the world. Including Father Gachet, who came to Saint Lucia in 1938, Father Coupperie, appointed bishop of Iraq in 1819, Father La Touche in Quebec in 1734, and Father Perroy, bishop of Burma in 1921, a long line of missionaries from the Vendée had spread the Gospel around the world and had, and still have today, extraordinary lives. Perhaps they had all read of the extraordinary adventures of Father Labat in the islands.

Beyond the admirable humanitarian work they did for the population as a whole, both priests were particularly devoted to the Indian population still surviving on Dominica. I think that our visit did them some good, because they could talk a little about the old country, but for me it was a moment of happiness and peace. We had been at sea for ten months now, and except for our stay on Barbuda, where we had made an entire congregation smile, we had not had such a warm welcome. Adventure for them was on the other side of every threshold they crossed. For us, it would continue on our next days at sea.

During our final week on the island, the teams switched. I went with Tim and Glen onto the slopes of the volcanoes. The one that was just above Roseau became the focus of our attention, because it presented an immediate danger to the town. The week before, the two researchers had already installed a satellite dish on the slopes, but they were now seeking a location closer to the promontory overlooking the town. The requirements for security and a clear space were met, but we were unable to find a rock set deeply and firmly enough in the ground. Glen then recalled an experiment that had produced good results on his expedition to Mexico. He decided to put a dish on the roof of one of the houses overlooking the town. Its owner, a Canadian banker, allowed us to come to his house to determine whether this was possible. The roof had a flat portion that was perfectly suitable. The house was built on deep foundations, fulfilling the final criterion. So our last installation was on a roof.

A few days later, the mission came to an end. A meeting was arranged with the head of the local government. More than fifteen installations had been measured and catalogued, and the countless hours of work and hundreds of kilometers traveled impressed the official. He asked Tim and Glen to send him the first results as soon as possible and agreed that we could return a year later in order to measure the differences in position of our installations, for we had left the pins in place at every location.

It was already late June. I realized that I would not have enough time to get to Trinidad to put the *Prince* in safe harbor before hurricane season. The place I had reserved would certainly be given to someone else. I would find a solution when I got to Martinique.

Some of the invaluable equipment was loaded onto a cargo ship headed for Miami, the rest shipped by plane.

On the way to the airport, Tim said, "We'll meet again in October in Miami, when we'll have some of the results from the expedition. As for Las Aves, Andy and Alberto will go with you again in 2002. Don't worry, I know you did all you could. I only think next time you'll need a bigger boat."

"I think so, too. As soon as I know anything, I'll send you an e-mail."

"What are your plans now?"

"We're going to Martinique to put the *Prince* in safe harbor for the hurricane season, and then I'm going to do an aerial survey of Carriacou."

"Oh, what's going on there?"

"I've been told that there's an underwater grotto with several drawings on the walls. I intend to find out whether that's true before I sail there early next year. After that, I'm returning to France to continue research on fish sounds in the La Rochelle aquarium until October, with a researcher from the French Maritime Institute."

"Sounds like a good plan. And then?"

"New York and Washington for a meeting with the Bacardi Foundation, then Cape Canaveral to see Professor Gilmore, Miami to see you, and then the preserve on Saint Barts to continue listening to fish. The year will be over before we know it. I have no time to lose in finding a new boat for another expedition to Las Aves."

"Good-bye and good luck."

"Bon voyage, and have a safe trip to Miami."

"Before I forget, for Dominica, mission accomplished."

The channel between Dominica and Martinique gave us a turbulent crossing. Once again the sea was raging, even though it was the season of tropical calm. We had to take the shortest route to get to the shelter of the leeward side of the island, which required us to sail into a southeast wind, losing us a lot of time. When we reached harbor, we were finally in calm water. I decided to moor off Saint-Pierre, which would enable me to reserve by telephone a place in dry dock at the Sailor's Marina and to visit the museum. Here, too, Mount Pelée had caused havoc when it erupted early in the morning of May 8, 1902, with thirty thousand victims. This was another reason to continue our work. One day, researchers may be successful in their quest for more precise predictions. It's only a question of time and resources.

Two days later, the *Prince* was swinging at the end of a crane. It would be in dry dock for a while. When it was laid up, I looked in the log. It had just covered without mishap twenty thousand miles from La Rochelle in ten months. It certainly deserved a little rest. So did I, but only for a few days.

11

The Cries of Fish and of Men

A cave on Carriacou — A volcano hidden underwater —
The aquarium — Quiet research — Tarpons and clown fish —
New York wounded — Washington and our sponsor — Return to
Cape Canaveral — Results of the work on Dominica —
The marine preserve on Saint Barts — Return to the source —
The Prince *speaks to me.*

Only two sounds reassured me slightly: the whistling of the air bubbles escaping from my regulator, following the rhythm of my breathing, and the metallic clang that my diving tanks made against the walls. The beam of light from the lamp guiding my progress through this narrow passage gave me little comfort, because it illuminated only a dark channel that seemed to lead nowhere. The two divers following me had made the courageous decision to let me go first.

On the boat, just before we went into the water, my two companions had informed me that this entry tunnel was no longer than twenty meters. With no further information, this distance seemed longer and longer as I moved forward. A rope that might have guided us in the darkness had never been put in, which was not very smart. The spot had to remain secret, since there was no way of installing security devices. Finally, after ten interminable minutes, the narrow tube seemed to open out. A vast mirror with dark reflections covered this new space. This must be the underside of the surface of the little lake ending up in this cave. I was at long last able to lift my head out of the water. The echo of the bubbles escaping from my regulator and breaking on the surface of the mirror filled the space. My two diving companions, Mike and Jürgen, surfaced a few seconds later. The cave, now illuminated by our three lamps, appeared in all its splendor. The vault must have been about ten meters high. A few scattered stalactites, like the dangerous teeth of a harrow, seemed to protect it. Ancient dampness seeped down the side walls. On the side opposite us stood a narrow promontory a few centimeters high grazing the surface. We took up position there to be able to better observe the interior of this mysterious cavern. In a heavy silence, our lamps swept over every possible space of the quiet cave. If we saw a drawing or a sculpture on the rock, like the ones I had seen on Barbuda and Saint Croix, our curiosity would be satisfied and we would forget our dangerous dive. Their discovery would provide a solid argument to come back in better conditions in 2002 on our next expedition, when we would have sophisticated photographic devices and better exploration equipment. But for the moment, the plan was not warranted, because there was no sign, no trace that men had ever given free rein to their creative imagination on the walls of this dark world. The sculptures apparently discovered by Jürgen, which were the reason for this exploratory dive, were nothing more or less than the fruit of his imagination. They were only vague shapes with no meaning, sculpted by nature over time.

I had flown to the island of Carriacou a few days earlier, after Raphael had returned to his family on Saint Martin. The *Prince* was in safe harbor. Thus, before continuing my listening in an unaccustomed place in France the following month, I could plan the next year's expedition. We already had Las Aves as our goal. If I found good reasons to come back to this cave, that would wind up the mission. But there were none.

And yet our casting off from the large old jetty of the little port of Hillsborough was encouraging for the future success of our discoveries. Among the local boats loading cows and chickens for transport to the market on Grenada, there was a gaily dressed crowd of onlookers watching us load our diving equipment onto our big fishing boat, convinced that we were off on a treasure hunt. Our destination had remained secret until the last minute, but rumors moved quickly among the huts and the old buildings of this charming island town. The rustling of the broad palms of the huge coconut trees lining the bay had whispered in the evening breeze that foreigners were about to make the island's fortune. Their casting off could not be missed. Our downcast faces when we returned later that morning were soon interpreted as a failure. The island of Carriacou would remain as it had been since the beginning — a place forgotten by everyone, even though a few smugglers still preserved its reputation among the Grenadines. The numerous bars could resume their conversations about a hopeless future. The quiet life would begin again tomorrow with the crowing of the roosters that still wandered through the old plantations converted into small farms.

The tour of the island I made with my diving friends was, on the other hand, very promising, because on the east coast between Point Kendeace and Point Saint Hilaire lies a huge untouched barrier of coral reefs. Protected from the open sea by the island of Little Martinique, it is inhabited by an enormous variety of coral fish and dozens of species of madrepore in good health. This was the only treasure we discovered here. But it was clearly worth a special trip warranting the use of our hydrophones sometime in the near future. Mike and Jürgen would make a more thorough exploration, or so they told me on the way to the Carriacou airport.

From the window of the Cessna that was taking me to Grenada for my flight back to Europe, I could see the large coconut grove on the south of the island as we banked, and then, still at low altitude, I observed bubbles breaking in the middle of the blue surface of the channel dividing Grenada from Carriacou. Hidden beneath the indigo sea, a short distance from Round Island, a new volcano was rising from the depths. Even though it was still underwater, it had already been named Kick-'Em-Jenny. It was still in the Leeward Islands, but not for long. The tectonic plate would continue its ineluctable movement and would soon force the dome of the volcano above the surface. As I left Grenada, I hoped that this upsurge from the bowels of the earth would take place without distress or tragedy, and then fell asleep on the flight that would take me from the warm tropical night back to old Europe.

I was in La Rochelle a few weeks later, in late July 2001.

Pierre was waiting for me at the aquarium with my listening equipment. We had agreed before I left La Rochelle a year before to take advantage of my experience to make recordings in the various tanks. Was the behavior of the species living here modified by their confinement in glass tanks? Had they preserved, as in their natural setting, their instinct to communicate? These were the reasons for the sound studies that lasted for more than a month. Roselyne, the manager of the aquarium, was fascinated by the research and had gladly welcomed me into her team.

It had been agreed that a certain number of conditions had to be fulfilled in order to carry out this work. First of all, we had to take advantage of the calm of the evening, after the visitors had left. Then we had to stop the air circulation systems in all the tanks, which meant turning off the many mixing and filtering pumps so that their noise would not interfere. Two species would be the focus of our research: tarpons and clown fish. For the former, I wanted to compensate for our misadventure on Vieques, where the Coast Guard had kept me from going to Salina del Sur Bay on the east of the island to record them. For the latter, the opportunity was marvelous. Like every tank in this gigantic aquarium, the clown-fish tank was carefully presented. In it I found all the flora to which these fish were accustomed. Their ecosystem, including the salinity and temperature of the water, had been respected to the letter, and there were various kinds of sea anemones and underwater plants common in their environment. Some couples had deposited their eggs on a rock and followed one another in ventilating their future offspring with their fins.

With the help of Pierre, the biologist and scientific director of the aquarium, a hydrophone was for the first time quietly slipped beneath the surface in the tarpons' tank. In the calm of the evening, seated on the steps facing the tank with our headphones on, we were waiting for a miracle. Jean-Paul, a marine ecology researcher from the Centre Nationale de la Recherche Scientifique-Institut Français de Recherche pour L'Exploitation de la Mer in L'Houmeau, who was interested in our experiment, had told me a few days earlier that our microphones had to be reliable enough and adjusted to the proper frequency to capture sounds emitted by the tarpons.

The large tank contained four of them, each about fifty centimeters in length. Their sleek, slender bodies covered with thick scales swam peacefully back and forth. The microphone, set ten centimeters beneath the surface, did not seem to interest them at all. Suddenly, as I moved closer to the glass to film them, one of them rose to the surface, took a breath of air, dove down, and turned to face me. Its body twisted two or three times, then it opened its mouth and with great effort emitted a preliminary groan. The sound was clearly recorded on our machine. The experience was repeated each time I approached the glass wall. The next few evenings, several of us observed them and filmed their behavior. Although we did not want to interpret their attitude as yet, we were certain that the sounds they emitted indicated that they had a means of communication. This discovery, for the first time in the world as far as we knew, could be the beginning of a study of their life and behavior. The recording of the sounds would probably not have been possible in the open ocean, because tarpons are very alert and move rapidly. They have seldom been observed because they are difficult to detect underwater. They swim in schools, generally feed at night, and flee at the slightest alarm.

Our second experiment was with the clown fish. We were not the first. Others before us had studied several species that emitted sounds. In most cases, these sounds effectively aid in their survival and successful mating. Our objective was to provide a description of the sounds produced by a particular species, the *akallopisos*, to defend their territory. We spent day after day in front of the tank recording their sounds. For each one, we tried to observe whether it was the sign of a specific behavior. The time allowed for listening in the evening was no longer sufficient. We then decided to vary the times and also listen during the day. Had you visited this wonderful aquarium during those times, you would have witnessed our production. We had installed two tiny speakers outside the tank. Watching the stage with a customarily silent glass curtain, you would have heard often incomprehensible dialogues between two madmen absorbed in their work. Their remarks certainly would have astonished you.

"It's the smaller one, a nasty character, attacking the larger one. He's the one who made the burst of tapping noises."

"No, I think it's the female. With her series of loud booms, she's telling him that he's invaded her territory."

The analysis of acoustic energy by means of the software for processing signals, producing oscillograms and frequency spectrums from the digitized signals, indicated by their acoustic characteristics that sounds of the *akallopisos* for intimidation were very clearly distinguishable from those of other species in the aquarium.

There was indeed a difference in sonic intensity between displays of intimidation by couples and individuals without territory. Even though the experiment had to be repeated and the sounds isolated, it was clear to us that there were exchanges that seemed to be succinct dialogues. The study resulted in a scientific paper written by Jean-Paul and published in the *Annales* of the La Rochelle Science Society in March 2003.

Between two recording sessions, I had the good fortune to find a new boat, the *Betelgeuse*, that fulfilled all the requirements for our future mission to the island of Las Aves.

In the late morning of September 11, while we were in the midst of recording, still trying to understand the cries of the fish, Pierre came to get me to watch television.

A plane had just crashed into one of the towers of the World Trade Center in New York. What we at first took for a frightful accident soon turned out to be a veritable attack when a second plane hit the other tower. The ensuing panic in the nearby streets, all shown live on screen, made our research seem quite insignificant. Other cries came through the screen, the death rattles of men and women.

I got to New York a week later, almost alone on the Virgin flight from London. The few other passengers were all afraid of another attack and were ready to defend themselves fiercely if our plane was turned into another flying bomb.

I had an appointment in Washington at the Bicardi

Foundation's headquarters to present the budget for the following year's expedition and to explain our research program for 2002 to the president, Mr. O'Brien. New York was the only possible landing site, since all other East Coast airports were closed.

I did not recognize Manhattan. There was practically no one on the streets, and the rare passersby seemed to have shrunk within themselves. The strange silence on the streets expressed people's inability to understand the shock of the barbaric attack. Broadway, usually swarming with people, had been completely depopulated to facilitate the constant comings and goings of the fire trucks with their screaming sirens. The entrances of firehouses were hidden behind mountains of flowers, and rows of candles with delicate flames lined the sidewalks. The department stores on Fifth Avenue, completely empty, had been changed into temples of nonconsumption. Sadness drifted above the city like a shroud. There had been almost three thousand dead, all civilians. All I felt was a sense of rage. How could this have happened? Who was so angry at America? Why were all those innocent lives lost?

On the train taking me to Washington, I hoped that from the heap of steel beams and blood that the towers of Manhattan had become would arise wisdom and understanding among people, but the scar was too deep. The world was quickly divided between two universal elements: law and force.

The capital had also been attacked; part of the Pentagon had been destroyed by another suicide flight. The same syndrome had overcome Washington. No one wanted to take the Metro. The White House was surrounded by huge cement blocks. The army was everywhere on the major thoroughfares.

It was in this atmosphere of sadness and confusion that I met again with Mr. O'Brien, the president of the Bacardi Foundation. He was waiting for me at home with his family. Obviously, the first subject of conversation was not our research and our adventures during the first year at sea, but rather the consequences of the tragedy for the future of the country. It was not until the next day, at dinner with his family, that he congratulated me for our work, about which he had received regular reports. The economic uncertainty for the coming months led him, as I could well imagine, to be cautious. His foundation would continue to support us, but only through June 2002. This was already quite generous on his part, considering the extraordinary situation of the country. He promised to come to Guadeloupe when the expedition was to set out for Las Aves.

Two days later, I met with Professor Gilmore at Cape Canaveral. The space program had been temporarily suspended. There would be no launches until further notice. I told him of our program for 2002 and the continuation of our quest for underwater sounds. When I told him of the difficulties we had had at sea and the method we had developed for getting good results, he again told me how pleased he was. He had received the numerous minidisks containing the sounds of various species of fish, and over the coming weeks he would begin deciphering the most recent ones. Preliminary results were more than encouraging, because in some cases he said he had heard and analyzed sounds the existence of which he had not previously suspected. I showed him some unusual photographs from our research in the mangrove forest on Barbuda and explained to him our good fortune in finding our "studio" on the island of Virgin Gorda. To add a little color, I also told him of our visit to the church in Codrington and the pastor's reaction.

"You see, Gilles," he said, "listening to fish sometimes helps you to understand humans."

That evening, while we were eating a delicious dinner prepared by his wife, the discussion took a more serious turn. I learned that the first experiments on the sounds of fish went back to 1965. Researchers in Miami had already developed a dictionary of sounds emitted by dragonets. Each sound was associated with a specific activity, such as eating, courtship, and fleeing danger. Dr. Steinberg of Miami had done the best work associating sounds with activities. He had catalogued six for food, three for movement, and three for pain. Underwater television had been the principal research tool at the time. But Professor Myberg, whom I had the good fortune to meet in Miami so that he could identify the mouth-to-mouth photograph I had taken off Las Aves, was the one who moved this new science forward. He replayed the sounds already recorded through amplified speakers to sharks while varying the frequencies. As soon as low-frequency sounds, below 800 Hz, were emitted — resembling those produced by a fish in distress — sharks rushed in from all directions.

"I am in the line of those researchers," Professor Gilmore told me. "The only difference is that we are going to apply these discoveries to the conquest of space."

"Suppose one day," I said, to bring things back to earth, "we could determine with certainty — through a satellite surveillance system — the sounds made by groupers when they spawn. Could we protect those areas and hence the groupers?"

"It's under consideration," he answered. "I've already alerted international authorities to the legal aspects of the plan."

Gilmore again urged me to be careful in my expedition and not to take needless risks during my recordings.

During the long drive to Miami, I asked myself even more questions. Were we listening to species in the process of going extinct, or were we recording the earliest primitive languages of a species still evolving? What would we not give today to have a recording of the last cries of the Neanderthals and the first sounds of the more sophisticated language of *Homo sapiens*? One species disappeared to leave room for the next . . . I was soon at the entrance to the marine university of Key Biscayne. In one of its laboratories equipped with many measuring devices and computers, Tim, as calm as ever, was waiting for me.

"The Caribbean plate has slightly changed direction." Those were his first words.

"Really? That's what our work on Dominica shows?"

"Yes, in part; we already had that feeling before going there, but this confirms it. Of course, we'll have to go back and make measurements on Las Aves, but the direction in which it is now moving was confirmed by the installations of our satellite dishes on the island."

"And what do you think the consequences will be?" I asked.

"We have to be very cautious, we can't draw any hasty

conclusions, but the fact that the plate has turned northeast rather than east is of sufficient importance for us to pay more attention to the northern islands of the archipelago."

"Like Saint Kitts and Nevis, for example."

"Exactly."

I remembered our arrival in darkness, after forty hours at the tiller, in the calm leeward bay beneath the volcano on Nevis.

"I think I've found a good boat for our expedition to Las Aves. Ninety tons, four cabins, a spacious hold. We'll sail directly from Guadeloupe to the island."

"Around when?"

"Early in the second quarter of 2002; I'll confirm the date by e-mail."

"Okay, have a good trip back to Europe; I'll wait to hear from you. We did good work on Dominica. Really good."

Three days later, I would be handing in my reports on the coral reefs to James, director of the coral surveillance program for NOAA. He had already been informed of our work from the various documents I had express-mailed to him.

"I didn't think you would explore so many of them," he said.

"It wasn't always easy. What most surprised us was the destruction caused by the elements"

"With global warming, things are not going to get any better."

I told him of my meeting with Richard on Saint Thomas. "Do you know him?" I asked.

"Not personally, but I'm following his work on this new virus with interest."

"When do you think you'll follow up?"

"It's very worrying. I'm waiting for a detailed report before sending a research team to his area. Right now, I'm reviewing the latest results of our satellite survey of the Caribbean arc, which gives us regular reports on surface water temperatures."

I told him of our planned visit to Las Aves and our hope of exploring the coral reef surrounding it.

"Talk to Richard; he may ask you to bring him some samples. Congratulations on your work; we could use more volunteers like you. By the way, thanks for the postcards from Vendée. I gave them to my father. You remember that my name is Hendee. My father is convinced that his ancestors are from there and that a spelling mistake changed the V to an H."

Another one, I thought. The *Prince* was certainly always right — the men from Vendée were everywhere.

The plane that took me to Saint Martin was empty. Not a soul on board. The American airline companies were going to have serious problems. René, a businessman friend who put me up, was already complaining about the absence of tourists. He asked me to give a lecture to his favorite club before I went on to Saint Barts.

The docks of Gustavia were deserted. Franciane, in charge of the island's nature preserve, was waiting for me in her little office on the port. As a result of extensive uncontrolled fishing, the underwater depths had been quickly depopulated and the sea grass damaged by the anchors of countless yachts. As a result, in 1996, the local authorities created the preserve, which is now divided into five regulated zones covering more than twelve hundred hectares around the island. Three are reserved for moderate fishing by professionals — one for mating and supervised fishing, and the last is completely protected. In all zones there is a prohibition against underwater fishing, throwing rubbish overboard, the collection of shellfish, and fishing with nets. Boats must use specially designated buoys for mooring. The personnel of the preserve have two boats for regular rounds through the zones. Various scientists — not enough to suit Franciane — come to conduct studies of the forty-two species of coral that have been catalogued and the restoration of the aquatic fauna partially destroyed in the past by underwater fishermen. We had her authorization for our recordings.

While carrying out a filmed survey with the preserve's equipment, I was surprised to note the great extent to which these areas had been repopulated. I came across a huge number of infant groupers and all kinds of coral fish that had found here the calm necessary for mating. Various sea grasses had recovered strength and now seamlessly carpeted the bottom.

Our recording program went off without a hitch, although in spots there was frequent interference from the engine sounds of passing boats. The cooperation of the reserve personnel, who sometimes had to accommodate our odd hours for listening, was exemplary. It was a great pleasure working with them. The zones were known and surveyed; all we had to do was drop our hydrophones over the side, listen, and record. The only difference between Saint Martin and Barbuda was that here there was no surprise and no danger. This nature preserve completely fulfilled its stated goals. Multiplying this example on other islands would be indispensable over the coming years. I was led to believe that at the beginning there was strong opposition from some fishermen, but now, in light of the results, everyone was pleased that it was there.

Before I left, I had the pleasure of giving a lecture on our work, arranged by Marc Thézé, manager of the Guanahani Hotel, for his clients and friends on the island.

Hurricane season that year had been less severe than in preceding years. It was already the end of November, and their destructive force had spared the northern islands. The *Prince*, well protected on the platform of the shipyard of the Sailor's Marina on Martinique, could rest in peace.

A few weeks later, after a brief stopover on Antigua to meet with the authorities again and make preparations on Guadeloupe for our coming voyage to Las Aves, my Air France flight to Paris made a brief stop on Martinique. Through the window on takeoff, I could glimpse in the distance the outlines of the docks of the marina where the *Prince* was being kept. I waved to him. Half closing my eyes, drifting off to sleep, I had the impression that he answered:

"Greetings, Captain; it was a wonderful adventure and any boat would have been happy to go on it with you."

He did not yet know what I had in store for him

Great Navigators and Their Ships

Bartholomeu Dias (*ca* 1450–1500)	The *São Cristovão*, the São *Pantaleão*
Christopher Columbus (1451–1506)	The *Santa María*, the *Nina*, the *Pinta*, the *Marie-Galante*
Vasco da Gama (*ca* 1460–1524)	The *São Rafael*, the *São Gabriel*, the *São Miguel*
Sebastian Cabot (1476?–1555)	The *Mathew*
Ferdinand Magellan (*ca* 1480–1521)	The *Victoria*, the *San Antonio*, the *Concepción*, the *Santiago*, the *Trinidad*
Giovanni da Verrazzano (1485?–?1528)	*La Dauphine*
Jacques Cartier (1491–1557)	*La Grande Hermine*, *La Petite Hermine*, *L'Émerillon*

Sir Francis Drake (1540[or 1543]–96)	The *Pelican*, the *Golden Hind*
Jacques Jean David Nau a.k.a. François l'Olonnois (*ca* 1630–71)	*La Cacaoyère, La Poudrière,* *Le Saint-Jean*
Baron George Anson (1697–1762)	The *Centurion*
John Byron (1723–86)	The *Dolphin*, the *Tamar*
Samuel Wallis (1728–95)	The *Dolphin*, the *Swallow*
James Cook (1728–79)	The *Endeavour*, the *Resolution*, the *Adventure*
Louis-Antoine de Bougainville (1729–1811)	*La Boudeuse, L'Étoile*
Joseph Antoine Bruni d'Entrecasteaux (1739–93)	*La Recherche, L'Espérance*

Jean François de Galaup, Comte deLa Pérouse (1741–88)	*L'Astrolabe, La Boussole*
Louis Isidore Duperrey (1786–1863)	*La Coquille*
Jules Sébastien César Dumont d'Urville (1790–1842)	*L'Astrolabe, La Zélée*
Abel Aubert Dupetit-Thouars (1793–1864)	*La Vénus*
Charles Robert Darwin (1809–82)	The *Beagle*
Jean-Baptiste Charcot (1867–1936)	*Le Pourquoi Pas?*
Albert I of Monaco (1848–1922)	*L'Hirondelle, La Princesse Alice*

Acknowledgments

Prof. Tim Dixon, RSMAS, Miami
Dr. Alan Smith, UC Santa Barbara
Dr. Glen Mattioli, NSF
Prof. Frank Gilmore, DYNAMAC, Cape Canaveral
Alberto Lopez, University of Chicago
Jean-Paul Lagardère, IFREMER, La Rochelle
Pierre Morinière, La Rochelle Aquarium
François Beauducel, IPGP, Guadeloupe
Sylvie Leroy, CNRS, Paris
Prof. Arthur Myrberg, RSMAS
M. Cheminée, IPGP, Paris
Andy Hibbie, University of Puerto Rico

Joaquim Bacardi, Miami
Robert O'Brien, The Bacardi Family Foundation
The employees of Bacardi France and Bacardi Portugal
Christophe Maincourt, Richemont Group, Miami
The Conseil général of La Vendée
The Fisheries Council of Barbuda

RFO Guadeloupe
Angela Swafford, Discovery Channel

The television services of Antigua, Saint Croix, and Saint Thomas
FRANCE 3, La Rochelle
STUDIO 70, La Rochelle, and Yves, its chief editor
Pierre Fernandez, AFP, Paris
Claude Dubillot, *Sud-Ouest*, La Rochelle
Gilles Prévost, *La Betelgeuse*, La Rochelle
Yves Gaubert, *Ouest-France*, Marsilly
François Marot, *National Geographic* France
Cyrille Vanlerberghe, *Le Figaro*, Paris
Pierrette, *Le Journal de Saint-Barth*
Carla Mendez, *El Nuevo Dia*, Puerto Rico
Sylvie Girot and Véronique Corbel, *L'Essentiel*, La Rochelle

Franciane Gréaux, Marine preserve of Saint Barts
René Mathon, Lions Club, Saint Martin
Joel King and his hotel in Saint Thomas
Marc Thèze, Hotel Guanahani, Saint Barts
David Greenhaw, Bitter End Yacht Club, Virgin Gorda
Dominique Cabre, Jolly Harbour, Antigua
Jose Koquix, Diving, Inc., Miami
Pascale Gousselan, psychiatrist and poet, New York

Thanks to my son, Jérôme, and Tristan, his Prince de Vendée.

Special thanks to Annie from La Rochelle and all the crew members who served on the *Prince de Vendée.*

ACKNOWLEDGMENTS

A special thought for our fishermen friends of Barbuda — John, Elvin, Joseph, Moose, and Bungy — for their efficient and courageous support. Your Rockhind will not forget your lesson in life.

Thanks to the *Prince de Vendée*, my constant companion.